Mind over Matter

Mind over Matter:

The Necessity of Metaphysics in a Material World

Brian M. Rossiter
and
Wayne D. Rossiter

ATHANATOS
PUBLISHING GROUP

Mind over Matter:
   The Necessity of Metaphysics in a Material World

by Brian M. Rossiter and Wayne D. Rossiter

Copyright © 2015 by Brian M. Rossiter and Wayne D. Rossiter

Published by Athanatos Publishing Group
   www.athanatosministries.org

ISBN: 978-1-936830-80-0

Quotations of the Old and New Testament are from the HOLY BIBLE, NEW INTERNATIONAL VERSION®. NIV® Copyright © 1973, 1978, 1984 by International Bible Society. Used by permission of Zondervan. All rights reserved.

Table of Contents

1	Introduction	1
2	Arguments from Science	10
3	Arguments from Philosophy and Logic	46
4	Arguments from within Theism	65
5	A Partial Synthesis	72
6	Appendix	83
7	Bibliography	93

Acknowledgments

First, we would like to extend thanks to Casey Luskin for providing insightful feedback on the original manuscript. We would like to extend a special thanks to Anthony Horvath and Athanatos Ministries for publishing the book and for working with us to develop the finished product. We thank Jeremy Finch, Quincy Hathaway and Corey Rugh for reading the original manuscript and giving us helpful suggestions from different vantage points. I (BR) also want to thank my wife, Jessi, for allowing me the time and lending me the support needed to create this book. I (WR) thank the numerous apologists on the ground for their hard work and support (you know who you are). I also thank my family and friends for their support, especially my wife, Melissa.

Mind over Matter:
The Necessity of Metaphysics in a Material World

Introduction

In 2008, John Lennox (a mathematician) and Richard Dawkins (a zoologist) met to debate the topic, "Has science buried God?"[1] Five years later, physicist Lawrence Krauss and philosopher William Lane Craig debated precisely the same topic.[2] In fact, the fields of geology, paleontology, mathematics, philosophy, evolutionary biology, cosmology, history, and a good many other "ologies", have found their way into this great discussion. And make no mistake, the fact that trained professionals from all of these academic backgrounds are meeting, writing, and rallying their respective sides tells us something very important. Chiefly, it tells us that all of these thinkers believe that their fields of study have something to say about both the existence of God and the metaphysical implications that would necessarily arise from Him. Many feel the existence of God can be falsified based on arguments from science and philosophy. As Peter Atkins once put it,

> "I have to set myself an honest target, which is nothing less than complete explanation. Nothing fudged. Nothing forgotten. The atheist argument fails, if in the end, it turns out that the universe had to be designed. It fails if any aspect of it had to be made. . . . The atheist argument begins to corrode if there are aspects of the human condition that science cannot touch."[3]

For others, the very same subjects are thought to provide strong foundational evidence for the existence of God. The general idea is that, if a creator God exists, then His hand should be imprinted upon something that science (in the broadest usage of the term) can report.[4] So as not to fall into a classic trap, consider the fact that a physicist who believes physics has something to say about the existence or nonexistence of God does not call his phys-

---

[1] Dawkins, "Has Science Buried God?"
[2] Krauss, "Life, the Universe and Nothing."
[3] Atkins, "What is the Evidence?"
[4] Though most in the theistic evolution camp reject this notion.

ics a religion. Given this observation, it is immediately clear that theists working in these various fields are not practicing religion, but are arguing that their respective vocations can potentially contribute evidences for or against one. We might do well to identify the fact that, when atheists use evidences from their fields of study to make a case against the existence of God, they do not consider themselves to be acting on religious motives, but when a theist does similarly in defense of God, they are accused of religious bias. And so, we (BR and WR) reject the premise that "creation science" and Intelligent Design (ID) are just examples of religious creationism "dressed in a cheap tuxedo" of science. As we will discuss shortly, science (among other fields) seems to be very much in the business of informing us about the utility or uselessness of the God Hypothesis.

An old and overused saying is that there are two things we should never discuss: religion and politics. The justification for this view is that broaching such topics will lead to heated quarrels and the deterioration of relationships. The point is not that those two items must be discussed, but that there is a very particular reason why they are not. These topics strike at the core of our being. The age of political correctness has taught us to avoid "offending" others at all costs, and apparently disagreeing with others on these touchy topics is very offensive. But, if in avoiding uncomfortable conversations, we are left talking with others about hockey and cheese curls, what kind of relationships do we really have?[5] Henry David Thoreau once griped of the increasing inanity and superficiality of public discourse, writing, "We are eager to tunnel under the Atlantic and bring the old world some weeks nearer to the new; but perchance the first news that will leak through into the broad flapping American ear will be that Princess Adelaide has the whooping cough."[6] This is precisely where we are today. Anyone who pays attention to popular culture will see that we have willfully exchanged intellectual meat and potatoes for brightly colored pixie sticks and lollipops.[7] We have largely decided to avoid dis-

---

[5] Not that we have anything against hockey or cheese curls.
[6] Thoreau, *Walden*, 36.
[7] For example, I (WR) recently saw an article posted on Facebook that read, "Stop Taking a Stand on Facebook". Irony notwithstanding (apparently the au-

cussing the meaningful things of life, and to occupy ourselves with the most meaningless ones.

But not all have assumed the supine position of ignoring the weighty topics of life. In specific, the issue of God has been front-and-center in public discourse, from the halls of Oxford, to the pub down the street, to social networking outlets like Facebook. On this front, one side of the ledger has been relentless and full-bore in advancing its perspectives. Anyone with their ear to the ground in this arena will concede that the loudest and most aggressive advances are attributable to the so-called New Atheists and secular humanists (the members of which are largely overlapping).[8] Folks under these banners have decided that playing nice is unnecessary, and the tone of discourse ceased being colloquial some time ago. In a recent article in *The Atlantic*, Emma Green writes,

> "[T]he tone taken toward organized religion, especially recently, has been more shout-y than shrug-y. At the 2012 Reason Rally in Washington, D.C., for example, 'a band fired up the crowd with a rousing song that lampooned the belief in 'Jesus coming again,' mixing it with sexual innuendo,' ... Attendees sported T-Shirts and signs with slogans like 'I prefer facts' and 'religion is like a penis.'"[9]

As reported, one of the major impetuses at this rally was to "ridicule" faith. William Lane Craig has offered that, "ridicule, mockery, and insult are their *modus operandi*."[10]

Today, we see public espousal of views like those of Dawkins, who believes that, "faith is one of the world's great evils, comparable to the smallpox virus but harder to eradicate."[11] Thusly, he believes that raising a child up in any faith is child abuse.[12]

---

thor was taking a stand on Facebook not to take stands on Facebook), it illustrates the social push to avoid serious public discourse in favor of posting what meal we chose for lunch today.

[8] Other common terms are "freethinkers", "Nones" and "skeptics."

[9] Green, "The Origins of Aggressive Atheism". We would point out that conferences founded on advancing "reason" typically show scant signs of its usage.

[10] Craig, "The Origins of Aggressive Atheism."

[11] Dawkins, "Is Science a Religion?"

[12] Cooper, "Forcing Religion on Your Children."

Jerry Coyne (professor of biology at the University of Chicago) recently wrote an article titled "How to get rid of religion", and argued that, "the religiosity of a country is highly correlated with the dysfunctionality of the society; that is, the more dysfunctional a society, the more religious it is."[13] The aforementioned Krauss recently offered, "Journalists like to say that there are two sides to every story, but the great thing about science is that somebody is usually wrong…[saying that] religion and science are rivals is like saying that Bambi and Godzilla are rivals."[14] Speaking of journalists, the late Christopher Hitchens wrote a book titled, *God is Not Great: How Religion Poisons Everything*. Perhaps the most strident, Peter Boghossian (a philosophy professor at Portland State University) has written a book, *A Manual for Creating Atheists*, in which he happily considers himself an "evangelical atheist", wanting to produce, "legions of people who view interactions with the faithful as clinical interventions designed to disabuse them of their faith."[15] His tactics are brutish and abrasive, and he encourages his followers to engage the faithful in debate in their homes, synagogues, and places of work. In his book, he lays out a multi-step plan for erasing religion from the public sphere, which involves "stigmatiz[ing] faith-based claims like racist claims", "financially crippling" faith's purveyors, getting faith listed as a mental disorder, and "treat[ing] faith as a public health crisis."[16]

Those who do not share the view that faith is a great evil should take these examples seriously. The tactics of Dawkins, Boghossian and the rest are working. For example, a recent episode of NPR's *Morning Edition* covered the rise of the "Nones", which are people who self-report as having no faith whatsoever. Referencing the most recent findings from the Pew Research Center, the show reported that this subgroup of the citizenry has increased from 2 percent in 1950, to 16 percent in 2010. The last five years have seen geometric growth in this group. More than 7.5 million Americans have abandoned their faith since 2012, and the proportion of the population self-identifying as Nones present-

---

[13] Coyne, "How to get rid of Religion."
[14] Ibid.
[15] Boghossian, *A Manual for Creating Atheists*, 19.
[16] Ibid. 212-221.

ly (in 2015) is 21 percent.[17] Even among the remaining faithful, the body has been softened to the point of incoherence, and naturalism is the prevailing view. Society increasingly accepts the idea that science is the only way to know truths about reality, that science only deals with naturalistic explanations, and that all other claims are things you must believe on blind faith, in direct opposition to facts and evidence. As a case in point, one of the most popular arguments against all forms of religion is that "faith is believing something you know ain't true."[18]

The faithful have buy-in-large bought into the sentiment. For example, Denis Lamoureux (a Christian) has argued that "Science deals with the physical world . . . . establish[ing] laws to describe *what* nature is made of and *how* it works. . . In contrast, religion (and philosophy) . . . deals with ultimate reality behind or beyond the physical world."[19] Broadly speaking, this is the view espoused by theistic evolution, in which naturalism is sufficient to explain *everything* in the universe (and often even the universe itself). The laws and processes derived from naturalism are entirely sufficient. No God is needed to directly explain *what* and *how*. Thus, religious or metaphysical claims are entirely separable from the explanations science has to offer us about the world we live in (or even our internal experiences as biological creatures). A firewall has been built between metaphysical "beliefs" and empirical evidence. But, we believe that this demand to accept full-on naturalism is baseless, and its arguments philosophically and logically untenable. Our view is that naturalism is winning in the public sphere, not based on merit, but because of crafty rhetoric and tactical implementation.[20]

This book is not an apologetics treatise on "defending the faith"—though both of us are Christians. However, in final analy-

---

[17] Tobin, "Americans have 'Lost Their Religion.'"
[18] This originates from Mark Twain, but is incredibly popular today among atheists.
[19] Lamoureaux, *I Love Jesus & Accept Evolution*, 11.
[20] It is important to state at the outset that we are obviously not denying the use of naturalistic explanations to describe or explain some (perhaps many) phenomena, but instead reject the notion that *all* phenomena can be explained naturalistically.

sis, it is a defense of the existence of God. It is a defense on the many fronts that are most active in the public sphere, and thus incorporates arguments from science, philosophy, logic, history, and many other areas of thought. The purpose of this book is to arm people with the knowledge, logical arguments, and debate tactics needed to combat methodological and philosophical naturalism in situations where their encroachment is in error.[21] Rather than ignoring the discussion, or returning rhetoric with rhetoric, we have produced a short, tactical guide to addressing the most common arguments philosophical and methodological naturalists use to supplant supernaturalism (or any metaphysics that assumes nonphysical action). Peter Boghossian argues for what he calls "street epistemologists", which are everyday people, trained and equipped to take rational thinking to the streets. In this book, we attempt to seriously engage the "on the street" discussions, and to meet arguments head on. Thus the responses we offer are succinct but highly effective, and user-friendly in that they should be accessible to just about anyone. Rather than getting too bogged down with jargon, or attempting to offer comprehensive discussions on each topic, we offer logical points that are intended to expose the core problem with each opposing argument, bypass messy and drawn-out conversations, and get at the heart of the matter.

Before proceeding, a disclaimer and clarification must be made. We do not want to give the impression that all of the terms we've introduced are somehow synonymous, nor that the topics discussed will neatly fall into categories of classification. Faith, religion, belief, theology, metaphysics, and the like are not the same things. Nor are naturalism, atheism, agnosticism, scientism, etcetera. However, all of these items are deeply connected. In reference to the beginning of this chapter, religious faiths often make claims about the nature of reality, just as certain scientific findings or theories extend into areas of faith, philosophy and metaphysics. For example, we will discuss naturalistic claims about the ontolo-

---

[21] Roughly defined, methodological naturalism is the idea that we operate (usually in a scientific sense) under the assumption of natural, physical or material causes (i.e., that supernatural, metaphysical or immaterial causes aren't acting). Philosophical naturalism goes a step further in holding that supernatural, nonphysical or immaterial causes do not exist.

gy (development) and teleology (goal-directedness) of physical systems like the universe, biological life, and evolution. Thus we will deal with the reasons for choosing either unguided processes or intelligent agency as explanations for particular phenomena. However, in doing so, we are not contending that a defense of faith requires something like Intelligent Design theory, or that all discussions of naturalism (say, in science) lead to theologically relevant conclusions. Our goal is simply to show that ascription to a completely naturalistic view of reality is unwarranted, and the failure of naturalism in some areas does have theological ramifications. For this reason, proponents of faith-based metaphysical claims should be able to apply many of our responses to the defense of their own belief systems. But before moving on to the arguments, a quick word about the nature of debate is in order.

Debating 101

Many shutter at the idea of debate. Still, truth be known, they are participants (even if just passively) to debates nearly every day. If they read a newspaper, they are hopefully getting factual news, but those facts are necessarily filtered through the opinion the journalist is trying to convey. Daniel Patrick Moynihan once said that we are welcome to have our own opinions, but not our own facts. But this statement is not always true. As a comical example, ESPN personality Matthew Berry demonstrated the power of truthful facts in the hands of an opinionated reporter.[22] In his well-known "100 fantasy facts" article (which he produces prior to the start of each NFL season), he warns that "all we do is give opinions. Oh, we disguise it as facts, as truths, as scouting X's and O's, but they really are all just opinions." He then proceeds to make a compelling case as to why you shouldn't pick Indianapolis Colts' quarterback Andrew Luck—who happens to be the 2nd best fantasy quarterback in the league—in this year's fantasy draft. His point is that facts can be used to make many cases, including those wildly misdirected. If one wants to justify a prior bias, sufficient facts are usually available. This of course is dangerous, and it is likely a significant source of the ships-passing-at-night dialogues

---

[22] Berry, "100 Fantasy Football Facts for 2015."

that take place in the public. Most of us come to the table with partial facts, pseudofacts, and even well-constructed logical arguments extending from those facts. So, why can't we all come to agreement? First, there are the unavoidable prior biases. Along with these biases, there is typically a vested interest in the belief, and something to be lost if we give it up. Sometimes that loss is simply egotistical. But other times, it is of true value to a person's worldview or sense of self.

Perhaps one of the most frustrating aspects of dialoguing on important issues is that sometimes a false belief wins out over truth, simply because of superior style and tactical placement. Unfortunately, style is every bit as important as substance, and the individual who can *sound* rational and intelligent (even humorous or supremely confident) has an unfair advantage over those lacking these qualities. This is something we can all work on. Likewise, wit, brevity, clarity and response time all play a role in how effective we are in conversing. The goal is to clearly communicate facts or ideas as succinctly as possible, and to be able to respond to counter claims in a reasonable and convincing manner. We should avoid ambiguities in terms or phrasings, being precise and direct. Getting caught off guard is the best way to lose a debate (even if the facts are on your side). Being wrong about the facts is another good way to lose a debate. Often, one can lose credibility in the larger discussion by simply being shown to be wrong on a small ancillary fact. We can work on all of these things as well. The best way to improve our ability to dialogue about views and beliefs is to engage topics in controlled or safe environments. I (WR) am a bit of a nut in this regard. I debate myself in the shower, when driving, or walking the dog. Every time I hear something that strikes me as incorrect, I start planning out how best to show it. I listen and read the debates of others, and my library is roughly a 50:50 split between those arguing for my beliefs and those arguing against them. I traffic skeptics groups, atheistic and freethinking websites, and even follow relevant players on social networks. As I often tell my students, it's good to know what you don't believe. That is, it's good to know another person's views as well (or better) than they do.

In terms of debate tactics, there is nothing wrong or immoral about wanting to succeed in a discussion. Remember, in many cases the person who is arguing in opposition to your view is the least likely person to change his or her mind about it. Often, engaging in conversations where views are in conflict is not for the individuals involved, but instead for those who are listening in. Aside from the benefit of feeling confident that your own views are true, this is the real value of debates and dialogue in these areas. Neither of us have ever witnessed or been part of a debate in which a debater publicly conceded being wrong. But, we have both seen many situations in which the validity of one view over another persuaded many onlookers. Most importantly—and admittedly most difficultly—debate really shouldn't be about sport. These things really aren't games, and there is no gain in winning on a technicality, when you know your view is wrong. At the risk of losing credibility, it is important to be honest in discussion, and admit where your view is shown to be failing or inconsistent (or even just incomplete). Dialogues often stop being debates and start being genuine conversations when such concessions are admitted, and honesty should work to your benefit in most cases, if for no other reason than all involved identify you as reasonable and open-minded.

Chapter 2: Arguments from Science

Argument:
"We can replace God with science."

Examples:
"Take God out of the equation. . . . what are we going to put in that blank? Well, how about science?"[23] – Jim Holt

"The divine spark that caused life in the first place is not so 'divine'. Everything we know about biochemistry and the discovery of complex amino acids in space tells us that it's quite likely that the complex organic molecules on which life is based either arose in space or developed on earth."[24] – Lawrence Krauss

Response:
"How exactly does science create or cause anything?"

Why it works:

We won't spend a great deal of time discussing this particular argument from atheism. On its face, it is completely nonsensical. The statement suggests that science can stand in place of God as an explanation for things. Of course, it does not logically follow that, because our logical framework of scientific thinking can describe phenomena, God therefore doesn't exist. It doesn't even necessarily mean that God wasn't involved or wasn't necessary. Having said that, there can be no mistake that some large (potentially complete) set of phenomena can be explained by science, and do not require God (at least directly). So, there is a partial truth to the argument.

That is why our response works. It cuts away the major thrust of the opponent's charge, and reveals a gaping logical misstep in the argument. Science is not an object or person. It is not a physical or material thing. It is a logical construct (an epistemology actually) that deals in the abstract currency of ideas and thoughts. Science describes patterns and phenomena, and attempts to ex-

---

[23] Holt, "Why Does the Universe Exist?"
[24] Krauss, "Life, the Universe, and Nothing."

plain how they work. But science has no causal power. Science cannot bring something into being. Science *can* render equations that describe the behavior of the universe but, as Stephen Hawking once asked, "What is it that breathes fire into the equations and makes a universe for them to describe?"[25] For example, the law(s) of gravity existed before there was such a thing as science to describe it (them). So, a proper rephrasing of the argument would be that we can replace agent causation (God) with event causation (physical laws and matter).

That leads to a more useful discussion. I (BR) refer to this scenario as the "no baker needed" fallacy. Essentially, the claim is that science (or the physical things it describes) can replace God as an explanation for all phenomena. The fallacy here is to equate *describing* things with *causing* them or *creating* them. Take a donut for example: Imagine that you are shopping at the grocery and are asked to participate in a taste-test for a new pastry in the bakery. Loving pastries, you agree. The bakery staff hands you a pastry, and gives you a few minutes to sample it and make some mental notes about the product. You observe that there are strong notes of cinnamon, and perhaps a trace of clove. Being that it is a pastry, you surmise that it is probably largely made up of flour, butter, sugar, and some type of leavening agent. Judging by its texture, it was almost certainly fried in oil. All in all, the "mystery pastry" appears to be a particular type of donut. When you complete your assessment of the product, the staff returns and asks you what you thought about it. You might proceed to accurately describe the ingredients, the reasons why each ingredient was necessary, how those ingredients relate to one another, and why the frying process was necessary. But you would not conclude from your taste-test that the donut made itself.

While this example may seem silly, it perfectly fits the absurdity of the analogous claim that science can replace God. Simply describing *that* something happens in the natural world—such as the fact that complex organic molecules arose on earth, for example—tells us absolutely nothing about where those molecules came from or how they were ordered in such a way as to produce

---

[25] Hawking, *Brief History of Time*, 174.

life. Donuts do not make themselves, and there is scant evidence that complex information-loaded biological systems do either.[26]

As a final thought, the form of the argument presented in Lawrence Krauss's quote presents another glaring problem. Typically, the move to replace God with science is not based on the demonstration that science *has* explained all of the phenomena, nor their ontology. It is based on the assumption that science *can* explain everything we once needed God for. Krauss (in the quote above) concedes that we don't know how life arose, or even if it happened on this planet, as opposed to some other planet. He's flatly wrong when he suggests that we're anywhere near a plausible mechanism for getting from chemistry (organic molecules) to biology (a first cell). But for most who use the argument, demonstration of these things is unnecessary. The very fact that one or more possible naturalistic explanations can be conjured is evidence enough to dismiss God. We deal with this logical misstep under the argument "We don't need to resort to God when we have other possible naturalistic explanations", and we refer you to that discussion.

Argument:
"We have the fossils, we win."

Examples:
"I wish the view I heard expressed by Lewis Black, the stand-up comedian, was more widely shared. He said he won't even debate evolution's detractors because 'we've got the fossils. We win.'"[27] - Sean Carroll

---

[26] We're not going to deal with the issue of life's origins in this present conversation, but we do point to, among many others, the recent comment from famed physicist (and agnostic) Paul Davies that "the spontaneous generation of life by random molecular shuffling is a ludicrously improbable event." (*The Cosmic Blueprint*, 118).
[27] Carroll, *Forms most Beautiful*, 10.

"Brown University biologist Kenneth R. Miller likes to sum up his views about evolution with a bumper sticker slogan: 'We have the fossils. We win.'"[28]

Response:

"We all have the fossils. Our theories are supposed to explain them."

Why it works:

This is one of those power moves that attempts to begin the game on your goal line. Rather than actually debate the efficacy of Darwin's mechanism in explaining the patterns of life we see, the opponent assumes the very patterns that need explaining, and uses them as evidence for the mechanism. This is circular logic at its finest. To illustrate the problem, suppose a man asks his friend why he likes a particular football team. His friend replies, "Because they're good." The man then asks why his friend thinks the football team is good, to which his friend replies, "Because I like them." Thus, both claims require that the other first be true. This is circular reasoning. Such circularities are commonplace in evolutionary biology.[29] If a theory is attempting to explain data (fossils), the data cannot be the explanation for themselves.

Thus, the tactic here is to demonstrate that we all have access to the same patterns and data, and must produce explanations for them. This is an important lesson for the uninitiated. When an evolutionist shows us a pattern in the fossil record or a phylogeny showing similarities and dissimilarities among taxa (species), this is not proof of Darwinian evolution. This is the pattern Darwinian evolution is supposed to explain. So the question becomes, which model of evolution or creation is able to most comprehensively explain the patterns we see? For more on this, see our discussion of Darwinian evolution as a model under the argument "ID isn't testable."[30]

---

[28] Harman, "Scientist puts Faith in Evolution."
[29] For example, homologous traits as evidence for common ancestry, but ancestry must be known to determine if a trait is homologous.
[30] There are also many good books on this topic, which survey the existing literature in the arena of evolutionary biology. We suggest Stephen Meyer's *Dar-*

As a secondary consideration specific to the Judeo-Christian faith, there is an unspoken assumption that the very existence of the fossil record is somehow incompatible with this brand of theism. Unfortunately, a similar view is often held by proponents of these faiths. For example, I (WR) recently spoke to a group of Christian college students on behalf of a *Ratio Christi* chapter. During the Q&A, a student asked how the Christian could possibly reconcile the fact that 99 percent of the species that have existed on the planet are extinct. This seemed to indicate "a lousy designer".[31] What most Christians (we cannot speak on behalf of Judaism) don't understand is that *all* forms of Christian theology contain an evolutionary process consistent with this pattern (though not always the Darwinian form of evolution). For example, most "Young Earth Creation" (YEC) views affirm that God created organisms according to their kind. Thus, these views do not argue that all species were directly made by God. Instead, the "kinds" directly created by God would go on to speciate *within* their kind over time. Further, most YECs also believe that death comes into the world through the Fall of man (furthermore, there is extinction associated with a global flood). Thus, the discovery of extinct species that are markedly different from their ancestors is entirely expected on the YEC view. Similar explanations exist for the Old Earth or "Day-Age" view of creation, whereby God still creates "kinds" of organisms, but large epochs of time pass, during which death and the capacity for microevolutionary change are already active in the creation. Additionally, it could be argued that, given the existence of dynamic environments, it would be poor design if organisms *weren't* able to change over time. To summarize once more for clarity, there is no Christian creation story that fails to imply some form of evolution compatible with species' extinctions.

---

win's Doubt, Fodor and Piattelli-Palmarini's *What Darwin got Wrong*, and, in shameless self-promotion, *Shadow of Oz*, by Wayne Rossiter.

[31] We deal with this under the topic of "Argument from poor design."

Argument:
"Intelligent Design commits a god-of-the-gaps fallacy."

Examples:
"As soon as we start highlighting specific places where we think we glimpse God's handiwork, we open ourselves to the old 'God of the gaps' problem."[32] –Karl Giberson

"A more serious problem with progressive creation is that it is a God-of-the-gaps model of origins. . . . The difficulty with this understanding of divine action is that when physical processes are discovered to explain a gap once thought to be a site of God's action, His purported intervention vanishes in the advancing light of science."[33] – Denis Lamoureux

"In the absence of having any clue as to what the explanation might be for something, just settle for saying 'a god did it'. This is the 'god of the gaps' move."[34] – A.C. Grayling

"Currently, intelligent design (ID) is a type of God of the gaps argument."[35] – Peter Boghossian

Response:
"Is there any situation in which it is fair to invoke God?"

Why it works:
This is perhaps the most common and nefarious argument against all forms of theism in which God is active in His creation. The basic foundation for the argument is that the supernatural activity of God has been invoked time and time again, only to be replaced by naturalistic explanations. Thus, as science has grown, God's job description has gotten smaller. Given this pattern, the

---
[32] Giberson, *Saving Darwin*, 216.
[33] Lamoureux, *I Love Jesus & Accept Evolution*, 25.
[34] Grayling, *The God Argument*, 74.
[35] Boghossian, *A Manual for Creating Atheists*, 172.

atheist (or theistic evolutionist) feels that theists are just looking for remaining holes in the scientific enterprise, and plugging them up with God, until science explains them as well. While there is evidence that this is not entirely the case, the response we've offered here cuts to the heart of the debate. The god-of-the-gaps argument is itself fallacious. It's a logical trap. It holds that, since we have a scientific explanation for a thing, God didn't do it (at least not directly).[36] It also holds that, even if science has not explained a thing, we are not allowed to insert God as an explanation, lest we have to later replace God with science again. That is, whether or not science has an explanation for a thing, God (or any non-natural cause) is not permitted. This is akin to playing a game where the rules are, heads I win, tails you lose. It's rigged.

This also assumes that science has the ability to explain all things that are "real" (i.e., physical), and assumes philosophical materialism or naturalism. For example, Michael Shermer concedes that, "In science we have a certain assumption. It's called 'methodological naturalism'. That is, we assume that there are natural explanations for natural phenomena."[37] Yet, scientists use naturalism-of-the-gaps routinely. Can't explain what caused the universe to come into existence? No problem. Simply invoke self-causation or multi-verse theories. Can't explain the origins of life? No problem, just invoke panspermia, or the RNA-first theory, or the "backs of crystals" theory, or biological "front loading." It doesn't matter if the hypothesis is true, or even tenable, so long as it assumes naturalism. Thus, our response points out that they have assumed the very thing that's up for conversation. An Intelligent Design theoretic is not some cockamamie stop-gap, like putting bubble gum in a crack in the dam. It is a defined methodology for discriminating between phenomena that arise from agent-based intelligent action and those that can be attributed to undirected mindless processes. As we mention in several places, the utility of

---

[36] In a real discussion, it's worth clarifying the difference between describing a pattern or process, and explaining it. This is particularly relevant to the problem of ontology and teleology, which we address in our "No Baker Needed" argument. Explaining how or why something exists is different than explaining that something exists.

[37] Shermer, "Debate on the Origins of Life."

these methods is not really in dispute when applied to engineering, cryptography, forensic science, behavioral ecology or archaeology (to name a few). It only gets labeled as "god-of-the-gaps" when the agent would necessarily be non-physical. Intelligence is not the damning factor from the atheist's perspective. The abandonment of naturalism is.

Argument:
> "We don't need to resort to God when we have other possible naturalistic explanations."

Examples:
> "I think you do get rid of an agent if the agent is superfluous to the explanation. . . . I mean, why bother when you've got a perfectly good explanation that doesn't involve guidance?"[38] –Richard Dawkins

> "It is intellectually lazy to just stop asking questions and stop looking for physical explanations and just say, 'God did it;' that's lazy."[39] – Lawrence Krauss

> "The implication of God as an explanation of anything is an admission of defeat and ignorance disguised as a pretense of understanding. I set myself a challenge that everything in and of the world, the body and the spirit can be understood without needing to invoke a god."[40] – Peter Atkins

Response:
> "Creating one or more just-so stories based on naturalism is quite different than demonstrating that God isn't a better explanation."

---

[38] Dawkins, "Has Science Buried God?".
[39] Krauss, "Life, the Universe, and Nothing."
[40] Atkins, "What is the Evidence."

Why it works:

This is tied to the assumptions made in the god-of-the-gaps argument. Namely, the opponent assumes that naturalistic hypotheses (no matter how ridiculous) are always superior to metaphysical ones—and more likely, that the latter are off limits entirely. To quote Sir Arthur Conan Doyle's famous character, Sherlock Holmes, "Once you have eliminated the impossible, whatever remains, however improbable, must be the truth." The assumption is that God is impossible as an explanation. For evidence of this, consider the famous (or infamous) words of Richard Lewontin,

> "It is not that the methods and institutions of science somehow compel us to accept a material explanation of the phenomenal world, but, on the contrary, that we are forced by our *a priori* adherence to material causes to create an apparatus of investigation and a set of concepts that produce material explanations, no matter how counter-intuitive, no matter how mystifying to the uninitiated. Moreover, that materialism is absolute, for we cannot allow a Divine Foot in the door."[41]

Now, let's consider an analogy: You're hiking with a group of scientists and your group happens upon a great chasm, deep ravine or canyon. You need to get to the other side. Thus, the question is "how do we get from here to there?" The scientists begin producing all of the possible scenarios or solutions. You could jump . . . but that's highly unlikely to work. All evidence seems to suggest that no human can leap the chasm, and it's not really close. You could crawl all the way down, and then back up. You could walk all the way around in one direction, and, if the chasm narrows there, you might be able to cross. Or, if the chasm shallows, you might stand a better chance of climbing down and back up again. Etcetera. The scientists can begin assigning probabilities to all of these possibilities. Some are more promising than others, but they're all on the table. It's important to note that all of these are unverified hypotheses, not demonstrated solutions. While they're working on the problem, you notice a bridge directly in front of

---

[41] Lewontin, Review of *The Demon-Haunted World*.

you. You say, "Why don't we just use the bridge?" They reply—to your astonishment—"No. we can't use the bridge. Somebody made that."

Of course, invoking agency in the design of nature is slightly more complicated. So, what if you're now part of a second group of hikers, years in the future, and you come up on the other side of the chasm? For the sake of the analogy, you see signs that in fact someone had been on the side you currently stand on, and that someone had been directly across the chasm at some point too. You ask "how did they get from there to here?" Ideally, the bridge would still be standing there. But suppose it's not. The only difference now is that, in the prior case, you were part of the party trying to figure out how (or even if) you could get to the other side. In the present case, you *know* that someone has gotten across, but you don't know how. Now you start to work through all of the ways to get from there to here. All of the same basic options are present. But, the option that may well have the most explanatory power is to assume a bridge was there. Then, the solution is quite easy. Clearly, you need a way to discriminate between blind processes and intelligent ones, which is precisely what ID theorists try to do.

This is not a perfect analogy, but it makes the point we're sure many of you see: methodological naturalists are looking for particular types of solutions to the problem, while disqualifying others from the discussion entirely. This is important. If a convincing case is made as to why a particular solution is lousy, methodological naturalists will typically just roll through their rolodex of solutions, and offer you the next one. The presumption is that they can do this *ad infinitum*, because new possible solutions can always be created. As physicist Sean Carroll explained it, "we can always concoct elaborate schemes to save the phenomena."[42] The ability to produce naturalistic explanations that could *possibly* be true functions like an endless series of red herrings, aimed at never allowing us to discuss the possibility of a non-naturalistic explanation for a phenomenon or pattern.

---

[42] Carroll, "Does the Universe Need God?"

Our task is not to completely and absolutely exclude all naturalistic solutions. This would be proving a universal negative, which is not feasible. Our goal is to postulate a model of causation that is both logically reasonable and also preferable to competing hypotheses in its explanatory power. Likewise, the scientific naturalist in no way gets off the hook by simply producing additional speculative scenarios. To produce additional naturalistic explanations is not to win the argument. Hypothesizing possible explanations is not the same as demonstrating real explanations. So long as we are allowing the possibility of positing causal power in metaphysical agents, it is clear that a model based on intelligent activity is more likely to be true. It only becomes impossible as a solution when we *first assume* only naturalistic explanations are possible.

Worse, if we pay close attention, we can typically see that the putative naturalistic explanations grow in improbability (i.e., the more they offer, the worse later explanations are). For example, current knowledge and experimentation suggests that the likelihood of a spontaneous transition from chemistry to biology (first life) is exceptionally small. Perhaps improbable to the point that it approaches impossible.[43] The inability to produce a demonstrable (or even potentially capable) mechanism for creating first life has left many naturalistic explanations in its wake (there are plenty of equally unlikely naturalistic explanations for how life could emerge on earth). Because they are so untenable, we are also considering various panspermia models, in which life is seeded here by (or from) an alien source. Some assume directed (i.e., intentional) introduction of life, while others posit accidental introduction aboard some meteorite or asteroid that strikes the earth. The noteworthy thing is that all of these panspermia theories are markedly less likely than getting an abiogenesis event on earth, because they rely on an abiogenesis event on a planet we cannot examine, *and then* survival of life through space, and survival of life upon impact on our planet. Assuming the evolution of hyper-intelligent life elsewhere is even *less* probable.

---

[43] For more discussion, see Horvath, *The Golden Rule of Epistemology*, 32-33.

But, isn't the postulation of a metaphysical cause as equally undemonstrated as the infinite line of naturalistic conjectures? Yes and no. Most feel that invoking a metaphysical cause (particularly if that cause is intelligent) likely seals us off from being able to produce a specific mechanism that describes how and why the agent causes the phenomenon. But, we think this leaps to a second step, and ignores the first.[44] The first step is to ask which explanation is more probable given the evidence in hand. While there are a plethora of specific causal explanations, they all fall into one of three kinds of explanations: physical necessity, physical chance or contingency, or intentional agency.

Something like first life seems not to be a physically necessary event. If only physical materials and processes are available to us, there are no existing principles that require life to arise. The apparent rarity of life in our neck of the universe, as well as its (assumed) single unique origin on this planet, strongly suggests that no deterministic mechanism is at play. Still, physical necessity is probably better than relying on chance in a purely physical system. The random or chance-based assembly of a single cell (or even a string of events from some replicating molecule to a single cell) is exceedingly improbable. All "evolutionary" hypotheses for first life suffer from both exorbitant improbability and from seeming impossibility, due to the interdependence of the parts. Without getting into the details, (and at the risk of using the dirty word in science), the cell is irreducibly complex. This is not to say that every single part is needed to get a cell, but that there appears to be a set of structures that are tightly interdependent, often forming recurrent (i.e., circular or self-referencing) systems. That is, there are interconnected structures in the cell that require each to exist before the other can, effectively creating gargantuan chicken-or-the-egg problems for explaining the cell in *toto*. It is not clear that *any* naturalistic explanation based on chance mechanisms is capable. While we might not go so far as to say, based on the interdependent structures, that chance-based models for the spontaneous

---

[44] We can—and do—come to the conclusion that intelligence is necessary to explain a phenomenon without having to know how or why the intelligence acted. That is a second question that extends from the first observation.

formation of a cell are logically impossible, we're comfortable in saying that they are exceedingly improbable. When a pattern or structure cannot be explained by physical necessity or chance, only one option remains live: intentional agency. As mentioned elsewhere, we accept this as patently obvious for structures made by the human mind (no scientist is attempting to explain the formation of the computer I'm typing on by means of physical necessity or chance). So, the question that must be asked is why have we decided that intentional agency must be excluded when considering biological systems? Indeed, such views are not excluded at present (i.e., directed pan-spermia), so long as the mind is itself reducible to explanation by naturalistic processes (note the circularity).

Argument:
"Highly improbable things happen all of the time"

Examples:
"We are all susceptible immediately to the idea that coincidences are special. Something that happens to us means a lot. And often that belief resides in the discrepancy between the laws of probability and the workings of cognition."[45] – Lawrence Krauss

"I am here. This is extremely improbable. Vanishingly small. Because if my father, when he was having intercourse with my mother, had coughed [sic] so different sperm had fertilized the egg, I wouldn't be here. My brother would be here talking, in fact. Not only that. If my father's father had coughed..."[46] - V.S. Ramachandran

Response:
"What do you mean by a 'highly improbable' event?"

---

[45] Krauss, "Life, the Universe, and Nothing."
[46] Ramachandran, "Beyond Belief: Session Ten."

Why it works:

This is a risky response, in that you're handing over the topic to the opponent. But, you're banking on him or her offering a false example that will work in your favor. For example, Eric Anderson recalls a story in which an atheist and an ID proponent were debating on a radio show.[47] In the debate, the atheist purportedly said, "what are the chances that you and I would be in this room, at this time, on this show, debating this topic?" If such an example is given, the follow-up response must be "Do you believe that happened by chance?" It is not a chance occurrence that the two men were debating. In fact, it is a strong demonstration of the power of agency to collapse probabilities, such that the event necessarily happens. The two men had chosen to become educated in fields that could apply to the topic of discussion, they had coordinated the event with the radio station, they had agreed on a time that would fit both of their schedules, and they had agreed on the topic they were going to discuss. Almost nothing in the scenario was left to chance.

The problem with what most people mean when they say "improbable things happen all the time" can be illustrated by another example. In 2009, the famed biologist Francisco Ayala (who also claims some form of theism) argued against the use of a "universal probability bound" in detecting intentional design. The logic of a probability bound is that some specified patterns or sequences can be sufficiently improbable to conclude that, even given all of the probabilistic resources of the entire universe, the event could never happen by chance.[48] To use an example, one could pixelate the Pablo Picasso painting *Guernica*, and then quantify the specific arrangement of the pixels in space. It's fair to say that randomly arranging pixels over and over again would be unlikely to reproduce the painting. As mentioned above, that's what intelligent agents are able to do. A skilled artist could look at Picasso's painting, and then, with just one or two attempts, establish an approximation of the painting that would outperform a random-painting generator. If the complexity (or informational con-

---

[47] Anderson, "Probability and Design."
[48] Dembski, *Design Inference*, 213.

tent) of the painting is sufficiently high, we can mathematically conclude that the painting could never be reproduced by a random process. Ayala takes issue with this view, by alluding to the argument that improbable things happen all of the time:

> "You don't need to argue with professor Dembski [who invokes the probability bound] because, by his own argumentation, he does not exist. His genetic makeup has a mother component and a father component. The mother during her life will produce about 500 eggs, each one with a different genetic composition. The father produces about 10^{12} sperm, each with a different genetic composition. The probability that he will get the two right compositions is 10^{-15}.... Going one generation back, the probability that Dr. Dembski exists, by his own argumentation, is 10^{-45}. If I go one more generation the probability that he exists is smaller than the reciprocal of the number of atoms in the universe. Those arguments have no seriousness for any serious scientist."[49]

The great error in Ayala's thinking is that he either ignorantly or intentionally misses the essential piece of Dembski's argument. Dembski explicitly describes any probability bound as a limit in probabilistic resources to randomly find a specified outcome. If we consider a series of ten digits in a code (ranging from 0-9 at each position), and then ask what the likelihood of getting *any* of the possible arrangements would be, the answer is one. We have not specified any particular outcome, but instead have simply asked for one of the possible outcomes. However, if we asked a random drawing to produce a specified outcome like 1783748325, the probability of getting it would be one in 10^{10}, or one in ten billion. If we drew a new randomly generated ten digit code once every second, it would take us (on average) about 2,740 years to get the correct target sequence of numbers.

As another way to illustrate the problem, suppose we had 10^{15} people, all playing the same lottery. What is the likelihood that,

---

[49] Ayala, "Is Intelligent Design Viable?"

upon drawing a winning number, somebody will win? Again, the answer is one. Somebody will win. Now, what is the probability that a particular individual would win? One in 10^{15}, or one in a trillion. If we were to randomly draw a new winning number every second, it would take approximately 274 million years for the individual to win. By analogy, if we randomly drew egg and sperm combinations from Dembski's parents once every second, it would take the same amount of time to get another individual with Dembski's DNA composition.[50] The critical oversight in Ayala's analogy is the absence of a specified outcome, and for this reason his argument fails entirely. There is a difference between specifying an outcome before the event takes place, and simply pointing to an outcome after the fact (post hoc). If there is no target specified *a priori* (no one was claiming that William Dembski's specific genotype was pre-ordained), then one cannot calculate the probability of an event. To not name a target outcome is to simply say that *something* will happen. Which, barring the end of space-time, the probability of *something* happening is one. So, the fallacy of using the "improbable things happen" argument is that it typically asks for any outcome, not a specified one.

Further, the argument for intelligent agency does not rely on improbability alone. In each of the examples, the specified outcomes reference other external items. An attempt to crack a ten digit code is an attempt to solve a code someone else has already specified. In a lottery, the argument pre-supposes that it is *my* number we're seeking prior to actually drawing the numbers. And, in the case of Pablo Picasso's painting, the master painter is trying to emulate a pattern already created. The most common example of the difference between producing any outcome and producing a specified one is that of Mount Rushmore. If we look at the face of any mountain, its topography is incredibly improbable. If we searched all of the mountains on all of the planets that have existed through the history of the universe, we would be unlikely to find one that had precisely the same topography of some given

---

[50] Here, Dembski's probability bound has purchase, because we are asking what the probability of getting a specified outcome would be, rather than simply viewing an event after it's happened.

mountain in the Himalayas. Yet, these highly improbable structures are the product of mindless processes. But Mount Rushmore is different. It too has an equally improbable topographic structure, but it does something more. It matches a specified pattern that is external to itself: it is a likening of the faces of previous American presidents. Nobody would look at Mount Nanga Parbat and suppose its structure was intelligently designed. But nobody would see Washington, Jefferson, Roosevelt, and Lincoln on the side of Mount Rushmore and doubt that an intelligent artificer was responsible.

Alternative form:
"We don't need God in order to explain complex systems in nature."

Example:
"The idea behind the complexity argument . . . is that just as the inner workings of a watch are too complicated to have arisen on their own, so too are the workings of the universe. . . . My response [is to] ask about tornados: 'Have you ever seen a tornado? Do you think that God has his finger on a button and just designs these incredibly intricate natural phenomena?'"[51] – Peter Boghossian

Response:
"What *does* explain that phenomenon?"

Why it works:
An argument like this is not really a serious argument. It's a rhetorical device that attempts to make the theist defend a straw man. Nobody argues that *all* complex phenomena are the result of God's direct hand. But worse, this type of argument opens up a regress that ends in precisely the place the atheist doesn't want to be. What does explain the creation of a complex tornado? First, it's worth noting that even the National Severe Storms Laboratory concedes that, "the process by which tornadoes form is not com-

---

[51] Boghossian, A *Manual for Creating Atheists*, 171.

pletely understood."[52] Unfortunately, the explanation stems from even more complex physical and climatological processes and forces. To explain one complex thing, you now need several *more* complex things. And this regress increases the informational requirement (one aspect of complexity) until you are stuck trying to explain the complexity of a universe that could give rise to galaxies, which could produce stars with orbiting planets like earth, which would have climates and topographies that permit tornadoes.

Bill Dembski has produced the "No Free Lunch" theorem to mathematically describe this problem. His observation that more information is needed to explain the creation of a phenomenon than is found in the phenomenon itself may well be a 4th Law of Thermodynamics (the "Conservation of Information"). That is, we can only explain the informational content of a thing (which is usually related to its complexity) by housing it in nested hierarchies of higher-order information-containing structures and processes. This applies to every system, including the biological machines found in the cell, and in the evolution of organisms themselves.

The second major response is to simply repudiate the claim. It is true that there are complex structures or systems that do not require explanations invoking intelligent causation directly. But many systems are in fact resistant to naturalistic explanations that do not include intelligence. There are numerous recursive systems at many scales of biology which are not just "irreducibly complex", but are logically impossible unless they exist in their entirety. The general form of a recurrent system is a situation where some product A leads to the creation of product B, and so on, such that $A \rightarrow B \rightarrow C \rightarrow D \rightarrow E \rightarrow A$. It is recurrent in that A is both the first cause *and* the final product of the series of events.

The obvious problem is that E is required in order to have A. How does one get E? Ultimately, from A. Thus, this is a circularity of precisely the same nature as the logical fallacy of circular reasoning. The first thing must exist in order to produce the last, and vice versa. Thus, neither can exist unless brought into exist-

---

[52] nssl.noaa.gov, "Frequently Asked Questions about Tornadoes."

ence by an outside force. A real-life example shows this sleight of hand at work. The Krebs cycle, for example, as the name implies is a circular system. After co-enzyme A (call this "*A*") binds to the 3-carbon pyruvate carrying an acetyl group (becoming acetyl CoA, call this "*B*"), it enters the Krebs cycle, binding to oxaloacetate (call this "*C*"). After many additional chemical transformations, oxaloacetate ("*C*") turns out to be the final product of the cycle as well (such that, *C*→*D*→*E*→*F*→*C*). So, the system begins and ends with oxaloacetate. But what's wrong here? First, you must have oxaloacetate to start the reaction, but you can't get it until the full sequence of reactions has already been completed. This is quintessentially the chicken or the egg. But it's much worse. You don't just need oxaloacetate. Recall, the acetyl CoA molecule is not from this system, but is the product of another set of reactions (glycolysis). Further, the process incorporates many energy carrying molecules (NAD+, ADP, and FAD) from elsewhere, which receive hydrogen ions from the system (effectively "loading" them with energy). This is the only reason the system exists: to produce energy. So, how then do we really get a reaction like this? We need an external source of acetyl CoA, oxaloacetate, and the energy-carrying molecules. This becomes a regress problem that only gets worse. Now we need other reactions to explain the Krebs cycle, and we need all of them to exist at the same time, or none of them serve a functional purpose. This observation is absolutely devastating. Rather than making a case for spontaneous order, systems like these make a case for the No Free Lunch theorem![53]

---

[53] It's also worth mentioning that, among the long list of such scenarios in biology, self-referencing circularity is a major barrier to all naturalistic theories on the origin of life. DNA is transcribed into RNA, which then negotiates translation of the code into a specified protein. But, proteins are needed for copying and reading DNA. As it turns out, you need RNA and proteins to make DNA, and DNA and RNA to make proteins. It's now clear that life cannot begin from proteins (because there is no path for moving from protein amino acid sequences to DNA coding for those sequences). It's also clear that information-rich DNA code cannot spontaneously arise. The hope currently rests in an RNA-first theory, but RNA code is derived from DNA code, and RNA is extremely fragile and decomposes rapidly. It seems that you need all three at the same time to get first life.

This is a good place to take issue with the suggestion that naturalism has provided a coherent (let alone plausible) mechanism for explaining complex systems. Here, we would like to focus on biological systems, because they are clearly different in kind than non-living examples of complexity, and they are by far the most complex. Specifically, we'll take issue with the Darwinian account for complexity. The Darwinian mechanism has at least one massive flaw: it destroys more than it creates. As Kirk Durston recently reminded us, "In the neo-Darwinian scenario for the origin and diversity of life, the digital functional information for life would have had to begin at zero. . . . An essential, falsifiable prediction of Darwinian theory, therefore, is that functional information must, on average, increase over time."[54] To talk about this, we advise using the following analogy (or something like it):

Suppose that you have a building that requires twenty steps on a ladder to reach the top. And suppose that you have one hundred ladders leaned against the building, each with a man standing halfway up the ladder. The goal is to get to the top of the building. In our analogy, each man represents an organism in a population, and each step up constitutes a mutation that increases fitness (while a step down equates to a loss of fitness).[55] Now, suppose that we will determine the movements of each man by a lottery system. With respect to mutations, the highest ratio of adaptive-to-maladaptive mutations ever recorded was roughly 1:8 (in that study, as in most, the majority of mutations were neutral, and can simply be ignored).[56] While the ratio has been shown to be lower in most studies, we're happy to take this very generous value. As applied to our analogy, suppose that each lottery drawing has a one-in-nine chance of allowing the man to take a step up, and an eight-in-nine chance of making him take a step down. Given this mechanism, how long would it take for any individual to reach the top? The chances of an individual getting up the ten steps to the top is something like one in one hundred billion. Note that we're being very charitable in this analogy. For example, if we asked the

---

[54] Durston, "An Essential Prediction."
[55] In Durston's scenario, we could call a step up some increase in information, and a step down would be a loss in informational content.
[56] Hall and Joseph, "High Frequency."

men to start at the bottom, their chances would drop to one in a thousand trillion.

Suffice it to say, it is much more likely that everybody will end up standing on the ground. That is, the net movement of all individuals will be down the ladder over time, not up. While individuals may achieve adaptive steps occasionally, on average, the process forces individuals down over time. I (BR) often call this "The Gambler", because it is much like a losing gambler who believes he can gamble his way out of debt. If, on average, he loses much more often than he wins, his debt will simply continue to grow. The lesson here is that a mechanism (like Darwinian evolution) which depends on chance mutations that disproportionately harm fitness will result in "genetic entropy" and the loss of fitness (and information) over time. As Stephen Meyer (among others) has observed, most evolutionists try to get out of this by simply assuming massive amounts of useful (adaptive) information before the mechanism gets started.[57] Even there, the net movement is towards degradation, not increased informational content, adaptation or structure. Basically, we still don't know how to create functional information-rich genes, which are the blueprints for biological structures. But that is precisely what both universal common ancestry and descent with modification requires!

Now, the evolutionist might object that each step up could be preserved by natural selection, making it permanent, and all progeny with steps down are eliminated from the contest.[58] But there is a problem with this view. It turns out that the *entire* set of ladders actually represents just *one* organism, because many mutations occur each time we copy our genome via cell division or reproduction. For example, while it is highly variable, each human is born with several dozen (perhaps more than one hundred) new mutations that their parents did not possess (and this is a conservative estimate). So long as there are always more negative mutations than positive ones (which is a consensus view in biology), then the net movement will be negative over time. We don't need to speculate or look to analogies. Science has amply shown this to be true.

---

[57] Chapter 11 of his book *Darwin's Doubt* is simply titled "Assume a Gene."
[58] This is basically Richard Dawkins's view, called "ratcheting."

Durston reports many such cases in studies of bacteria (where population sizes are massive and replication rates are incredibly short—both of which help the evolutionist's chances). A recent study by Fu et al. (2013) found that 73-86 percent of the rare (singleton) mutations in the human population are deleterious (correlating with known diseases).[59] Several similar empirically-derived studies now exist, and many modelling studies are giving us the same story.[60] In this sense, the classic neo-Darwinian mechanism is not capable of producing the observed patterns we see *in principle*. As an additional proof, if the theory was sufficient, we wouldn't need all of the alternative models now housed with the term "evolution". Note that we're not saying that evolution is therefore false, only that the Darwinian mechanism is clearly insufficient.

Argument:
"Intelligent Design is not science."

Examples:
"It is a simple matter of fact that intelligent design forms no part of contemporary science."[61] – Denis Alexander

"ID is not a scientific theory and should not be taught alongside the Theory of Evolution."[62] – Pennilyn (Penny) Higgins

"Intelligent Design is Not Science, and Should Not Join Evolution in the Classroom."[63] – Glenn Branch

"[L]et's be clear: Intelligent-design theory is not science."[64] – Michael Shermer

---

[59] Fu et al., "Analysis of 6,515 Exomes."
[60] To again quote physicist and agnostic Paul Davies, "One would suppose that random mutations in biology would tend to degrade, rather than enhance, the complex and intricate adaptedness of organisms. This is indeed the case, as direct experiment has shown: most mutations are harmful" (*The Cosmic Blueprint*, 109).
[61] Alexander, "Intelligent Design is Not Science."
[62] Higgins, "Why Intelligent Design (ID) is Not a Science."
[63] Branch, "Intelligent Design is Not a Science."

Response 1:
"Therefore what?"

Why it works:

This response seems snarky, but its goal is to force the individual to explain why it is important to label ID as a non-science. What does he or she think this distinction achieves? More often than not, this argument is simply an attempt to dismiss ID without actually evaluating its evidences and arguments—it is less an objection than it is a red herring. The assumption is usually that of scientism, in which the individual believes that science is the only way to know truths about reality, and thus ID (as a non-science) can be rejected out of hand. This response makes them say it outright. If this is in fact their angle, then you have now exposed a prior bias in their thinking, which has substantial costs to other views they probably hold. To engage those things, follow-up questions like "Is science the only way to know truths about reality?" or "If something isn't science, does it logically follow that it's not true?" may be in order.[65]

Alternative forms:

"ID is not testable." or "ID cannot produce testable predictions."

Examples:

"Scientists, including scientists who are Christians, do not use ID when they do science because it offers nothing in the way of testable hypotheses."[66] – Michael Shermer

---

[64] Shermer, "Not Intelligent & Surely Not Science."
[65] We point out that much of the discussion here would also have application to "creation science". Additionally, it's important to remember that the determination of whether or not something qualifies as science has no bearing on whether or not it's actually true. A thing can be true or false whether or not it is dubbed "scientific". Intelligent Design or Creation Science may well qualify as "science", in that the theories and hypotheses they create might be falsified. That is a different conversation.
[66] Ibid.

"A scientific theory is supported by extensive research and repeated experimentation and observation in the natural world. Unlike a true scientific theory, the existence of an 'intelligent' agent cannot be tested, nor is it falsifiable."[67] - Pennilyn (Penny) Higgins

"Intelligent Design Creationism is not science; it makes no testable predictions so it can not be falsified."[68] - John Cotton

Response:
"Do you believe ID is falsifiable?"

Why it works:
Perhaps the biggest objection made against the ID program is that it is supposedly not a genuine scientific enterprise. Specifically, the assertion is that ID is not capable of doing either of the two things that a good scientific theory needs to be able to do: it can neither make predictions nor be tested. Our response to this objection has two effects. First, it forces the opposition into a dichotomous choice: If ID is falsifiable, then it conforms to the most critical component of "science". Evidence can be brought to bear either in favor or against its claims. If ID is not falsifiable, then the individual can make no case against it, which begs the question why they think it to be false. The atheist rejects all claims that violate assumptions of naturalism. But, if a claim is not falsifiable, they have no justification for rejecting it. Their position should be pure agnosticism with respect to the claim.

The second thing this approach achieves is that it forces the opposition to begin to delimit what "science" means. In fact, if one desires a more drawn out—but often fruitful—discussion, he could simply ask for clarification: "What do you mean by science?" This presents a different dichotomy to the opposition. The argument that ID is not science assumes some quaint and superficial concept of science that depends on the scientific method, in which hypotheses produce predictions, which can be tested experimentally, and either accepted or rejected.

---

[67] Ibid.
[68] Cotton, "Intelligent Design Creationism is Not Science."

Both prongs of this argument need one point of qualification. We must also delimit what is meant by "evidence", and ask if it is overlapping with "reasons" for accepting or rejecting a claim. There are some things that can be demonstrated physically, as in the case of an experiment that attempts to manipulate the conditions involved in some physical phenomenon (say, the growth of bacterial colonies in different media). But other phenomena cannot be directly manipulated or reproduced (for example, the environmental conditions that immediately followed the extinction of the dinosaurs), and therefore can only be confirmed or rejected on the basis of logically evaluating competing hypotheses and the efficiency of their explanatory power (basically, the cost of increasing the complexity of the model against its ability to explain a phenomenon). On issues of describing historical events, it is not always possible to falsify all competing hypotheses, or even to confirm beyond a shadow of a doubt which hypothesis is most likely. The event itself, being a unique event in the past, cannot be recreated. We are left only with evidences of the processes putatively involved in its creation.

A classic example is to ask why Napoleon Bonaparte lost the battle of Waterloo. Was the loss due to overconfidence? Was it because the army was comprised of infantry, conscripts, loyalists and others, and thus lacked cohesion? Was it because the Emperor himself was plagued with medical ailments? Perhaps the grass was wet from prior rains. We will never really know, and it will be hard to determine which factors played a role in his defeat. Yet, it is a relatively well-established fact of history that it happened. So then, we must accept that we will not have a single "scientific" explanation that can be repeated, tested or falsified. Numerous areas of science are happy to accept these terms, and so it's not clear that, even if ID were not falsifiable, it would be disqualified as science. We continue that discussion in considering a more incendiary response:

Response 2:
> "Then, Darwinian evolution is also not science."

Why it works:
When it is said that ID does not provide a mechanism for making predictions, what is meant is that ID cannot tell us what we should expect to see in future events or patterns. Being that it could only tell us about the past (if it even does that), ID is chiefly a historical endeavor. We say "endeavor" because its critics certainly would not refer to it as a historical science, since it fails to qualify as a science in their minds. This assertion essentially means that ID provides no possible experimental aspects, because the design of an experiment would be predicated upon one or more predictions that could potentially be verified. Of what use would a theory be if it could neither be confirmed nor falsified? Of course, the very "design" of an experiment demonstrates the validity of "intelligent design" in its broadest sense. Thus, the real issue is not intelligence, nor design, but the idea that intelligent agency exists outside of purely physio-chemical brains in evolved material creatures. This, and a slew of other complaints against ID, turns out to be a philosophical assertion of the non-existence of metaphysical or supernatural (or *supra*natural) existence. As we'll see shortly, the atheist would (and has) gladly accept(ed) that many phenomena can be attributed to intelligent design, so long as the intelligent designer has naturalistic origins. If an intelligent designer can ultimately be reduced to explanation and description in purely physical terms, there is no problem.

At any rate, the objection that ID can't make predictions, and therefore is not science, is folly. Current conceptions of multiverse theory (or its converse, a self-caused universe) offer no predictive power about what we should see or what would happen in any other instance.[69] More frustrating, Darwinian evolution offers the

---

[69] For example, writing in the *New York Times*, Frank and Gleiser argue, "Today, our most ambitious science can seem at odds with the empirical methodology that has historically given the field its credibility . . . How are we to determine whether a theory is true if it cannot be validated experimentally?". Additionally, physicist Paul Steinhardt recently penned concerns in the journal *Nature*, saying, "[multiverse theory] is so flexible that it is immune to experi-

scientist no predictive power either. This will immediately skyrocket the blood pressure of many Darwinists, but it remains true. We will briefly outline the problem here:

As I (WR) have recently written,[70] suppose we wanted to apply the Darwinian theory of evolution to some bacterial colony in a stagnant pond in South Carolina. Now, what predictions would extend from evolutionary theory? What could we say about that bacterial strain going forward? Nothing. We have no idea when, where, how or if something might happen. It may go extinct tomorrow, or it may persist for ages. It may evolve into something strikingly different, or it may remain a bacterial strain forever. And the problem is not just that we lack knowledge of the contingencies involved (for example, knowing what selective pressures may arise gives us no more precision in making predictions). The underlying problem is that the mechanism itself is fundamentally indeterminate.

The biologist will object, claiming that evolution has been scientifically demonstrated in numerous lab (and real-world) settings. True, but this claim comes with some very important caveats. First, observing evolution after the fact is very different from predicting an evolutionary event before it happens (the prior has no predictive power). This alone eliminates most examples. Second, when someone like Richard Lenski or Francisco Ayala begins a study with a single selective force—artificially selected and enforced mind you—acting on lab colonies of bacteria, and ultimately elicit an evolutionary response (say, the ability to metabolize citrate[71] or antibiotic resist-ance[72]), the prediction takes on a probability function at best. No outcome is guaranteed or precisely repeatable. There is little to no precision in the predictions that might extend from Darwinian theory. We cannot know with certainty what (if any) adaptive solutions will arise, nor when. In extreme cases, we might be able to probabilistically predict *that* an

---

mental and observational tests . . . No experiment can rule out a theory that allows for all possible outcomes. Hence, the paradigm of inflation [multiverse theory] is unfalsifiable."

[70] Rossiter, *Shadow of Oz*, 148.
[71] Pennissi, "The Man Who Bottled Evolution."
[72] Ayala, *Darwin's Gift to Science*, 60.

event will eventually happen. By the way, this is by no means a criticism, and we do not mean to suggest that such frameworks are not science. Quite to the contrary, as we will now discuss.

The bulk of the case for Darwinian evolution is not in producing predictive models of the future, but is based upon the power of the theory in explaining the patterns of the past. In this sense, theories of biological evolution collectively represent what should be properly identified as a historical science. Darwin attempted to offer a mechanism that could explain the great chain of being (universal common descent).[73] His theory did not require this fact, but *assumed* it, and attempted to then explain it (i.e., his theory could still be valid, even if all life did not coalesce back to one original common ancestor). His theory also assumed great epochs of time, and the known patterns in the fossil record (though, as he acknowledged, the patterns were not easily explained by his theory). So, evolution—if by that we mean the diversification of life over time, extending from a common ancestor—is not proof of Darwin's theory. His theory *assumed* this pattern, and then attempted to explain how it could happen.[74] His basic argument was that natural selection would act to filter variation in form or behavior, preserving and amplifying adaptive traits and eliminating maladaptive ones. Combined with a mechanism for producing novel variation (mutations), Darwin had found a mechanism that, at least in theory, could produce the observed patterns of life. In short, Darwin was offering an explanation of history.

Here, a "testable prediction" would require an individual to predict certain facts or evidences in the past that have not yet been discovered, but should be, if the theory is true.[75] The best "predictions" that could emerge from this type of theory would be vague claims about the anticipation of transitional forms (which, on the whole, are largely not forthcoming), but no precise expectations can be formulated. The basic point is that the predictive power of the theory is roughly comparable to a theory that predicts there

---

[73] An assumption that is less secure today than at any time in the past century.
[74] See the argument "We have the fossils. We win." for more on this topic.
[75] Arguably, the most powerful proof of the theory is the independent discovery of genetic phylogenies that largely overlap with those based on morphology. We will also deal with this in a later section.

will be a car accident on New Year's Eve. It does not specify with any precision when, where, how, or even necessarily why the event will take place.

Returning to the argument, "ID is not science because it doesn't produce testable predictions," our job is simply to point out that, on that characterization of science, neither is Darwinian evolution. Further, none of the historical sciences would pass that test (sociology, anthropology, archaeology, ancient history, etc.). Finally, there is a second response to this argument. It's simply not true that ID is impotent in its capacity to make testable predictions. While small in number (but growing), consideration of biological systems in the context of intelligent agency has produced a few solid and testable predictions:

- While the first pass over the human genome led many biologists to believe 97 percent of it was "junk" (only about 3 percent of our genome codes for actual genes), ID proponents maintained that, if coded for by intelligence, it should all serve some purpose. Today, the ENCODE project has concluded that at least 80 percent of our genome is functional.
- The same is true for so-called pseudogenes, which were thought to be nonfunctional relics of evolution's past. Instead, modern molecular biology is showing us that "pseudogenes aren't pseudo anymore."[76] That is, the garbage heap has become a gold mine for functionality, including the direct usage of pseudogenes in regulating gene expression.
- Currently, there is a similar discussion about "silent" mutations (mutations that have no bearing on the organism, good or bad). Most of the genetic case for coalescent theory and the production of phylogenies is dependent upon shared silent mutations.[77] Naturalism argues that shared

---

[76] Wen et al., "Pseudo Genes are Not Pseudo Anymore."
[77] For example, if a historian of ancient civilization wanted to show that one text was derived from another pre-existing text (or that two existing texts were both derived from an older common text), the best way to do so would be to show shared typos (like the misspelling of John as "Johnn" at the same place in both texts).

mutations that are genuinely silent are best explained by a shared common ancestor. Shared ancestry is not necessarily in opposition to ID, but ID proponents have predicted that these mutations will turn out not to be so silent after all (that is, they will be functional). Over the past decade, molecular biologists have increasingly produced evidence supporting the ID prediction.[78]

So, when push comes to shove, ID has actually made predictions that can be verified or falsified by scientific findings. As one final ancillary, while neither of us find the evidence for Young Earth Creationism (YEC) convincing, there can be no doubt that many of its proponents in fact employ precisely the same "scientific" methodologies as their secular counterparts. As a great example, Nathaniel Jeanson (who incidentally holds a PhD in cell and developmental biology from Harvard University) recently published a study that tests the prediction of a young-earth time scale for mutation rates among animal kinds, using mitochondrial DNA sequences for 2,700 different species.[79] This is a peer-reviewed study using publicly available data, clearly defined predictions and methods, and thus is completely subject to validation or rejection, based on findings. As Stephen Meyer has pointed out in the past,[80] the claim that an intelligent design solution is not scientific is not only an attempt to throw sand in an opponent's eyes (the argument trivializes the debate by rendering it an issue of semantics), but it is also flatly false.

Argument:
"ID is not helpful to science, because agency does not lend itself to mechanistic description."

---

[78] See D'Onofrio and Abel (2014), Chamary and Hurst (2009) and Li, Oh and Weissman (2012) as a few examples.
[79] Jeanson, "Recent, Functionally Diverse Origin."
[80] Meyer, *Signature in the Cell,* 399.

Examples:

"Whatever processes were used to 'create' shut down on the sixth day of creation and are no longer a part of the natural order…It follows that there can really be no 'science' of origins, and we should not expect to understand the various mechanisms—all of them supernatural—that God used to create the world."[81] – Karl Giberson

"[Supposedly] natural selection cannot produce and assemble all these molecules. Intelligence can. How, exactly? If there is anything more than an empty word in this doctrine, we need, as before, to know how Intelligence is directed and what its power and limits are."[82] – Philip Kitcher

"The problem with intelligent design is that by saying 'the designer did it' you can explain anything."[83] – Kenneth Miller

"The problem is that by invoking something that is by definition not testable—there's no experiment we can run—you're off the page of science. So by involving 'extraterrestrial', 'supernatural', 'supra-natural', 'paranormal', something beyond the natural world, there's nothing we can do with that. It's kind of a conversation-stopper. It ends the discussion."[84] – Michael Shermer

Response:

"Does that mean agents don't exist?"

Why it works:

This objection is unfortunately identical to a fallacious argument often used by theists, and we should be careful to avoid falling into this trap ourselves. Many have said, "If we don't know

---

[81] Giberson, *Saving Darwin*, 4-5.
[82] Kitcher, *Living with Darwin*, 111.
[83] Miller, "How Evolution is Taught in the Classroom."
[84] Shermer, "The Origin of Life."

how it happened, how do we know *that* it happened?" This is potentially problematic. We routinely observe that things happen, without knowing why or how. For example, we may happen upon a dead insect on the sidewalk. We do not need to know how it died in order to know that it died. Similarly, the fact that a scientific model cannot mechanistically describe how or why agents operate in no way leads to the conclusion that agents aren't responsible for observed patterns or phenomena.

The grand mistake here is the assumption of a scientific schema that roughly traces back to Francis Bacon. Prior to his rendering of science, Aristotelian conceptions of science prevailed. Aristotle believed there were four types of causal explanations for phenomena: material, mechanical, efficient, and final (or ultimate). In brief, material causes refer to the physical matter under examination. For example, a block of marble. The mechanical cause refers to the actions (or events) causing the pattern under investigation. For example, the pick, chisel and hammer shaping the marble. The efficient cause refers to the reason that the mechanical cause is acting on the material one. For example, the chisel and hammer are being used by a sculptor to shape the marble. Finally, the ultimate cause refers to an external purpose for the entire system. Marble is being shaped by the sculptor using the pick, chisel and hammer, so as to produce a sculpture that is the likeness of Caesar. That is, the object under investigation (the sculpture) communicates a purpose by referencing another item (Caesar) external to itself. Under Bacon's recalibration of science, both efficient and ultimate causes are necessarily excluded from consideration.

So, the reason this response if effective is that it simply asks if the inability for science to apply either efficient or ultimate causes logically leads to the conclusion that the hypothesis is false. It could just as easily be said that the inability of science to detect (or consider) agency is not a proof of the non-existence of agents, but a failure of science to fully describe reality. In fact, many branches of science *do* permit agency as a causal mechanism. Just a few examples should suffice to make the point:

- No forensic scientist walks onto a crime scene, sees a man lying dead on the floor with four gunshot wounds and demands, "Explain how this man died of natural causes."
- The crucial—and still unsubstantiated—claim of Darwin was that natural selection could behave just as artificial selection (i.e., the intentional culling and sorting of variation by intelligent agents).
- The search for extraterrestrial life utilizes a methodology for filtering unintended or mindless patterns from those that would be created by intelligence for a purpose.
- For some time now, many origins of life biologists have been creating scientific research programs based on the idea that life might have been seeded here (either intentionally, as in the case of directed panspermia, or accidentally, in the case of lithospermia) by alien life.
- Archaeologists and anthropologists make similar dichotomous filtering methodologies in which artifacts can be categorized as having been formed by natural (unguided) processes or by guided processes for an ultimate purpose (say, tools or pots).
- The same is true for all research in behavioral ecology. For example, we are asked to discriminate between randomly assembled brush and bramble, and the intentional and teleological action of a ground-nesting bird that has created the structure.[85]

Shermer seems to be conceding this observation in the quote cited above. He unwittingly groups extraterrestrial intelligence with those things exterior to nature. This obviously is a misstep, given that scientists wake each morning to the hope that such life will be detected by our methods. Further, there is no reason to think that extraterrestrial life would be in some way unnatural. This presents Shermer—and like-minded individuals—with a serious problem. From the viewpoint of naturalism, there is little space between intelligent agency on our planet (i.e., human and

---

[85] One sub-branch of behavioral research in ecology is actually referred to as "theory of mind."

animal cognition) and alien forms of intelligence. If tomorrow biologists (in even larger numbers) came to believe that life on earth has intelligent extraterrestrial origins, what would Shermer do? Given that we still lack a formal explanation for the human mind and its creative capacities for acting, he seems to be appealing to the tautology of defining away agency. In fact, we suspect that Shermer is entirely open to the incorporation of intelligent agency into scientific explanations, so long as he feels the agents themselves might one day be reducible to natural causes. But that is a secondary step, entirely irrelevant to the question of whether or not intelligence is an adequate explanation for a phenomenon. The point here is that we often explain patterns or phenomena by invoking intelligent agency, so the fact that ID does so in no way discredits it as non-science.

There is an alternative form of this argument, which is that it is "intellectually lazy" to invoke a metaphysical cause. Particularly, when that cause is God, you often hear, "You think you have all of the answers, but atheism is comfortable with admitting we don't know." Presumably, as the previous quote from Michael Shermer suggested, invoking God as a cause is a conversation-stopper, because they don't feel that science can proceed after that. A few quick comments on these types of complaints: First, they're just complaints. They are not logical arguments against the existence of metaphysical causes. Second, formalized science emerged from within Christian civilization, mostly through Christian scientists, and their theism has never been shown to stop the scientific search. In fact, brilliant scientists like Newton, Keplar, Galileo, and Planck made world-changing discoveries while operating under the assumption that they were uncovering how God created the universe. Third, it is simply untrue that an individual who invokes a metaphysical cause believes he or she knows everything. Fourth, to that end, it is the metaphysician (in this case, the theist) who is not only conceding that there are limits to science and to human knowledge and understanding, but the methodological naturalist is being dishonest in their humility. While often claiming to "celebrate the unknowns,"[86] their underlying belief is that they *can know* all things (given enough time).

Argument:
"ID is creationism."

Examples:
"Intelligent design is nothing more than creationism in a cheap tuxedo."[87] –Leonard Krishtalka

"The creationism of William Jennings Bryan and the Scopes trial was a tragedy. The creationism of the Intelligent Design theorists is a farce."[88] –Michael Shermer

"As the case for intelligent design is elaborated, therefore, the position slides away from bare bones antiselectionism toward the religiously more evocative position of novelty creationism."[89] – Philip Kitcher

Response:
"Does it logically follow that it isn't true?"

Why it works:
The attempt to smear or besmirch ID as "creationism" assumes many premises we're not willing to accept. One might ask why binding ID to creationism is fruitful to the atheist (or even the theistic evolutionist). They're banking on the assumption that creationism is not scientific, but rather, religious, and therefore can be rejected out of hand. In their minds, creationism begins with the assumption of God (typically as literally described in Genesis), and then attempts to render a scientific description of the creation account. Put another way, it attempts to conform science to the standards of Scripture. Our response asks the more primary logical question, does it follow that such a view is necessarily wrong or flawed? The answer is no. It is entirely possible that an

---

[86] Ball, "Celebrate the Unknowns."
[87] Melott, "Two Views of Intelligent Design."
[88] Shermer, "ID Works in Mysterious Ways."
[89] Kitcher, *Living with Darwin*, 19.

individual could begin with theism, and then find scientific evidences for a creation event(s). For example, the biblical claim that the universe came into existence at a finite time in the past is currently accepted by most physicists and cosmologists because it has been evidenced by science. Even if ID was a form of creationism, so what? What follows from that observation? We would still have to ask if their scientific theories make sense of both their theology and the patterns we see in the physical world.

Of course, the second option is to flatly reject the proposition outright. It is not true that ID is creationism. One can find evidence for intelligent agency outside of humans, and still have no loyalties to a particular brand of metaphysics or religiosity. The "scientific" aspect of ID is that it is a methodology that attempts to determine when an inference to intelligent causation is justifiable, and when it is not. (Remember, science already does this on a daily basis). More importantly, many ID proponents maintain that it is science that has led them to theism, not the other way around. Thus, ascription to any "creation" story is a *conclusion* based on their science, not an *a priori* assumption.

Reaching a step beyond this, it is possible to reverse the tables. If it's a logical fallacy to assume a metaphysical view before doing the science, one could ask the atheist what assumptions of naturalism, materialism or physicalism they bring to the work bench before they do any science. If ID is just creationism, then perhaps methodological naturalism applied to science is just atheism. We actually feel neither is true, but the atheist is in no position to make such a claim.

Chapter 3: Arguments from Philosophy and Logic

Argument:
"You're an atheist about all the other gods. We just got one God further"

Examples:
"Dr. [William Lane] Craig doesn't believe in them; almost all of them, but one. So it's a difference of just one. He's an atheist about all of these other gods . . ."[90] –Lawrence Krauss

"We are all atheists about most of the gods that humanity has ever believed in. Some of us just go one god further."[91] – Richard Dawkins

"I contend we are both atheists, I just believe in one fewer god than you do. When you understand why you dismiss all the other possible gods, you will understand why I dismiss yours."[92] –Stephen F. Roberts

Response:
"That's like looking at a guy with no kids and saying he's basically a parent. The difference between one and none is infinite."

Why it works:

This has long been a favorite of the more devout atheists. Somehow, the fact that people reject certain notions or conceptions of God(s) leads to the conclusion that they should abandon all gods. While popular, it's among the most poorly thought out arguments in their tool box. The response we offer illustrates the major problem. If a God (any god) exists, that is literally the precise opposite of there being no god. Just as having one child makes you no less a parent than those with three children, rejecting the gods of Greek mythology in favor of monotheism does no damage to the metaphysical claim that an intelligent agent creates

---

[90] Krauss, "Life, The Universe and Nothing."
[91] Dawkins, *The God Delusion*, 19.
[92] Roberts, "Quote History."

the patterns we see in the universe. If their argument was valid, then the reciprocal would work as well: you could say, the difference between us is just one God, so why not slide our way (towards theism)?

Argument:
"Faith is believing something you know isn't true."

Examples:
"Faith is believing what you know ain't so" –Mark Twain

"Not everything that's a case of pretending to know things you don't know is a case of faith, but cases of faith are instances of pretending to know something you don't know."[93] –Peter Boghossian

Response:
"Could you please show me this definition in a dictionary or encyclopedia?"

Why it works:

This argument has recently become very popular among atheists. It's also one of the easiest to rebut, because it is so clearly rhetorical and illogical. For starters, the statement is self-contradicting. It is impossible to actually believe something you know is untrue. Belief obviously requires that you *believe* the thing to be true. Faith cannot, in principle, be something you know isn't so. By asking for some real example of this definition of faith, you call their bluff.

An alternative form of this type of argument is the attempt to establish faith and reason as antonyms ("faith is the opposite of reason"). But, it's quite clear that faith is in no way the opposite of reason. If something is not reasonable, then it fails some test of logic. The opposite of faith is doubt, skepticism, distrust, or disbelief, none of which are necessarily bound to the idea of reason.

---

[93] Boghossian, *A Manual for Creating Atheists*, 24.

Nothing more really needs to be said here. These are just rhetorical devices. They are not logically sound arguments.

Argument:
"Extraordinary claims require extraordinary evidence."

Examples:
"As Carl Sagan would have said, extraordinary claims require extraordinary evidence."[94] – Lawrence Krauss

"An extraordinary claim requires extraordinary proof."[95] - Marcello Truzzi

"Darwin's original claim of evolution by means of natural selection was an extraordinary claim in its time, so he was required to provide extraordinary evidence for it."[96] – Michael Shermer

Response:
"No. Extraordinary claims require sufficient evidence." Or, "If that's true, your views must have extraordinary evidence."

Why it works:

The first response is simply a factual correction. The attempt of the argument is to establish an impossible burden of proof for an impossible violation of nature. But, all that must be shown to establish a claim is that there is sufficient evidence for it. Typically, that would mean showing that the evidences (facts) are better explained by the claim being true than by it being false.

The second response is to immediately turn the tables. It's not as if the atheist or skeptic is in a safe position with respect to the evidence-claim relationship. What's at play here, as mentioned above, is the prior assumption that metaphysical claims are "extraordinary" and naturalistic claims are somehow all just ordinary.

---

[94] Krauss, "Life, the Universe, and Nothing."
[95] Truzzi, "On the Extraordinary: An Attempt at Clarification."
[96] Shermer, *Why Darwin Matters*, 50.

So what would qualify as an extraordinary event? Perhaps something that has happened once in the entire known history of the universe, and cannot be causally explained? Sure, we could again mention the unknown origin of the universe, its fine-tuning, or the origins of first life. It's also perfectly fair to talk about the numerous bizarre transitions in evolution, which also seem to have only happened once (by their worldview). The transition from prokaryotes to eukaryotes remains a profound enigma and only happened once. Numerous "biological Big Bangs" exist, and they each represent a unique transitional step, the explanations for which typically assume that conditions and dynamics were unique to those moments (thus explaining why we shouldn't expect to see them more than once). As Paul Davies has put it, "where a unique event is concerned, the distinction between a natural and a miraculous process evaporates."[97]

We could go even further and attack science's assumption of objective observation and the capacity for free-willed decisions based on the evidence. As wet computers, there is no way of explaining how the human brain is capable of creating a "self" (often called the Hard Problem of Consciousness by philosophers). There is no logical justification for why we should trust our own thoughts and convictions, if we are just "phenomenological glockenspiels".[98] There are many more such problems for their view, but these should suffice to demonstrate that they have a lot of heavy lifting to do. They hold many extraordinary claims that currently lack evidential support.

---

[97] Davies, *The Cosmic Blueprint*, 118.
[98] Harris, *The Moral Landscape*, 104. For an excellent discussion of this problem from a secular perspective, consider reading Thomas Nagel's book *Mind and Cosmos*.

Argument:
"If God acts in the physical world, He is rendered just another cause alongside all of the others."

Examples:
"Intelligent Design and scientific creationism seem inadequate to me, because they reduce God to one agent among other agents in natural history. If ID is true, then it implies that the agents of evolution are natural selection, sexual selection, God, mutation, chance, and whatever else you want to list."[99] –Karl Giberson

The direct action of God in the world is an, "unacceptable reduction of the Creator to an invisible cause among competing creaturely causes (making God just a physical interventionist poking an occasional divine finger into the processes of the universe)."[100] –John Polkinghorne

"Here's the danger of that from a theistic point of view, I think, is that whatever you find . . . — whatever 'God' is— it's just going to turn out that he is just an extraterrestrial intelligence. He cannot be outside of space and time."[101] – Michael Shermer

Response:
"Why does it follow that, because God interacts with His creation, He ceases to be its Creator?"

Why it works:

As an older man, our grandfather worked at a gas station, taking out the trash, cleaning toilets, and sweeping the parking lot. His friends became worried that he had fallen on hard times, and finally one of them approached him about it. "Why are you working as a janitor at a gas station?" asked the friend, to whom our grandfather replied, "Because I own the place." This argument is

---

[99] Giberson, *Saving Darwin*, 216.
[100] Polkinghorne, *Belief in an Age of Science*, 72.
[101] Shermer, *Audio Martini* podcast.

common (particularly among theistic evolutionists), but lacks any merit whatsoever. There is no logical reason why the activity of God in His creation would in any way limit His power as the ultimate Creator. In fact, this is precisely what the Christian faith holds in its claim that the Son of God became a human being (and a servant at that). So it is especially humorous that theistic evolutionists would have difficulties with a God that interacts with His creation.

Additionally, this objection tells us a great deal about the worldview of the person who is using it. They are espousing a purely naturalistic understanding of the world that negates, *in principle*, the possibility of non-physical agent intervention or causation. This, of course, is the very thing that is supposed to be debated! Can purely naturalistic causes actually explain the origin of both the universe and the biological world, as well as the high order of complexity within them? Or, does intelligent agency best explain these items? To again quote William Dembski,

> "Evolutionary biology allows for only one line of criticism. Namely, to show that a complex specified biological structure could not have evolved by any material mechanisms. In other words, so long as some unknown material mechanism might have evolved the structure in question, ID is proscribed. This renders evolutionary theory immune to discomformation in principle, because the universe of unknown material mechanisms can never be exhausted. Furthermore, the evolutionist has no burden of evidence. Instead the burden of evidence is shifted entirely to the evolution skeptic . . . [and] the skeptic must prove nothing less than the universal negative."[102]

Whenever this objection is raised, the person is directly telling you that they will not even allow the possibility of a non-naturalistic explanation; such a thing cannot exist under any circumstance.

---

[102] Dembski, "Blind Evolution or Intelligent Design?"

In this sense, it is eerily reminiscent of the "god of the gaps" fallacy. Not surprisingly, both arguments are often used in close proximity to one another. In both cases, God could not conceivably exist because He would be reduced to some type of naturalistic explanation and cease to be beyond the physical. On the one hand, if we can explain something naturalistically then there is no need to invoke a supernatural cause; if God were to act (i.e., actually be detectable), then He is no longer God. On the other hand, if we have no naturalistic explanation, then only one of two answers could suffice for the naturalist: the first would be to say that the event or phenomenon didn't happen to begin with (ala David Hume),[103] and the second would be to "wish on the future" and promise that we will one day uncover a naturalistic explanation for the event or phenomenon. Clearly, this entire objection is a trap because it presupposes that non-natural explanations cannot exist.

Argument:
"Who made God?"

Examples:
"[T]he postulate of a designer or creator only raises the unanswerable question of who designed the designer or created the creator."[104] – Christopher Hitchens

"Who caused God? [Theists offer] a prime example of the Fallacy of Passing the Buck: invoking God to solve some problem, but then leaving unanswered that

---

[103] Hume argued that, given the choice between accepting a violation of physical law or doubting the witness attesting to it, we should always doubt the witness's testimony, saying, "There must, therefore, be a uniform experience against every miraculous event, otherwise the event would not merit that appellation. And as a uniform experience amounts to a proof, there is here a direct and full *proof,* from the nature of the fact, against the existence of any miracle; nor can such a proof be destroyed, or the miracle rendered credible, but by an opposite proof, which is superior." (*An Enquiry Concerning Human Understanding. The Harvard Classics.* 1909–14.)

[104] Hitchens, *God is not Great,* 71.

very same problem when applied to God himself."[105]
– Rebecca Goldstein

"Why does God exist? . . . If God were to exist, he might wonder, 'I'm eternal. I'm all powerful. Where did I come from?'"[106] – Jim Holt

Response:

"If I write my name on a piece of paper, do you need to know where I came from in order to know I wrote it?"

Why it works:

When confronted with the overwhelming list of problems involved in attempting to explain the origin of the universe, the methodological naturalist will often press his or her opponent to explain God's existence. From their perspective, this objection provides a challenge that is equivalent to the one being posed to them; while the physicalist cannot account for the origin of the universe, the metaphysicalist cannot account for the origin of a Supreme Being, Creator or Intelligence. This is assumed to create a "tit for tat" stalemate in which neither side can prove its claims to be more plausible than the other. For example, in a brilliant lecture that is well worth a listen, physicist Paul Davies conceded,

> "To explain something, in the everyday sense of the word, means you have to start with something. You have to assume something as given. Something which you regard as self-evidently true, and so reasonable you don't question it. . . . The axioms of geometry represent a sort of levitating super-turtle that holds the rest up. The same general explanation applies when we come to the universe and the underlying laws. . . . Just as God can't explain God's self, the laws of physics or science

---

[105] Goldstein, "36 Arguments for the Existence of God."
[106] Holt, "Why does the Universe Exist?"

can't explain science. These are both levitating super-turtles."[107]

While Davies is perhaps more transparent than most, notice his willingness to call the debate a draw. This is a tacit admission that there are no *sufficient* naturalistic explanations on tap that are capable of explaining how the universe could have come into being. In this sense, metaphysical explanations cannot be disregarded as being inferior to purely naturalistic explanations of the problem. In fact, the move to ask where God came from effectively casts God and physics as equally reasonable explanations for the universe.

But it gets worse, of course. When confronted with the question of God's origin, we would do well to remind the individual that the origin of the universe, not God, is the question at hand. This is not to say that the question of God's origin presents no problem whatsoever for the metaphysician, if he is trying to explain the universe. Instead, the question is utterly irrelevant within the context of the present conversation. God's existence is a secondary issue to the origin of the universe. Something like a God or a designer could still be the best explanation for the origin of the universe apart from discussing the details of such an entity. As philosopher William Lane Craig puts it, "in order to recognize that an explanation is the best, you don't need to have an explanation of the explanation".[108] By way of analogy, this objection is something akin to saying that one must pinpoint the birthplace of the gunman in order to validate the victim's wounds. It simply doesn't follow that one must explain God's existence in order to establish that He is a plausible—we believe, the *most* plausible—explanation for the origin of the universe.

While this is a sufficient reason to dismiss the question about God's origin outright, there is yet another lethal objection to the claim. Though a number of philosophers and theologians (Craig being one of them) have already articulated this particular point, the problem remains that to ask where God comes from is to unavoidably enter into an infinite regress fallacy in which one could

---

[107] Davies, *Beyond Belief: Session 5*.
[108] Craig, "Who Created God?"

never get started with the objection to begin with. If an explanation for God's existence were offered, the methodological naturalist would simply move to asking the next logical question: who (or what) created *the thing* that created God? The same question could be asked *ad infinitum*. But we are not in an infinitely regressing state of existence. Additionally, the question "who or what created it?" only applies to things that come into being (as our universe did). It does not apply to things that are eternal, because they are uncaused. While the concepts of timelessness and eternality are probably beyond the human intellect's capacity for understanding, all parties are faced with this problem. However, theism has two advantages: First, the theist hits this boundary one step further along in the explanatory process (the theist has no origin for God, but does have an origin for the universe). Second, it seems more reasonable than not to believe that metaphysical objects (or beings) are eternal realities, as opposed to believing that physical or material objects are. Combined, these observations have typically taken the form of the Kalam Cosmological Argument (KCA), which essentially concludes that, because the universe began to exist and cannot be self-caused, its cause must be transcendent to the universe itself. However, we caution the classic usage of this argument—wherein it is posited that the cause of the universe must be an immaterial, timeless, and spaceless being—in trying to justify the existence of God. Because it is beyond the scope of the present discussion, we have included additional considerations of the KCA in the appendix.[109]

---

[109] We also deal with two related arguments, Leibniz's argument for existence, and the so-called "fine-tuning" argument.

Argument:
"Christianity was dreamed up by a bunch of Iron Age peasants."

Examples:
"The point is that our current understanding of nature has changed; we've learned things, it's changed and developed since the claims were made by iron-age peasants who didn't even know the earth orbited the sun; and therefore it's natural that science is inconsistent with those claims based on ignorance. And we shouldn't revere those ancient claims as sacred; they're ignorant."[110] – Lawrence Krauss

"Telephone Game as an analogy. . . is supposed to be to the long series of people that the Jesus story passed through from the alleged eyewitnesses, to those who repeated the story, to the authors of the Gospels, to the scribes that copied the Gospels that we have now. . . . Many of them are illiterate, Iron Age peasants. Some of them are suffering from bereavement hallucinations. Some of them are apparently having epileptic seizures complete with religious visions."[111] – Matt McCormick

Response:
"Funny, Christianity came after the Greek philosophers, and was advanced by kings, historians, trained intellectuals and even doctors."

Why it works:

This is an argument that is also growing in popularity (a clear result of the emerging coordination among the atheistic cyber-entities). But it's absurd on its face. Nobody thinks the Greek philosophers were ignorant, nor the Buddhist logicians, Roman historians or Arab mathematicians of antiquity. Christianity (and Judaism as well) was not created and advanced by mere fishermen alone. As the response clarifies, many kings, Caesars, historians,

---

[110] Krauss, "Life, the Universe, and Nothing."
[111] McCormick, "Amplification of Doubts."

theologically trained professionals, doctors, clerics, politicians, and a slew of others, advanced it. Further, this charge commits the so-called "genetic fallacy", in which an argument is rejected based on its source alone. Even if the religion was propagated by ignorant peasants, there's no more reason to reject their claims than to reject an eye-witness account to a murder because the testimony comes from an adulterer. So what?

Argument:
"Are things good because God says they are, or are they good, and therefore God acknowledges them as good?" (a.k.a. the "Euthyphro Dilemma")

Examples:
"Many people take it for granted that . . . divine commands, and only divine commands, could serve as the foundation for binding prescriptions on human lives and human conduct. Yet this general approach to ethical life has been in trouble ever since Plato's Socrates posed a dilemma for Euthy-phro."[112] – Philip Kitcher

"Plato had already recognized in his dialogue Euthyphro, that the statement 'God is good' is not a tautology, and that therefore the question of the nature and basis of goodness is a matter that can be discussed independently of theology."[113] – A.C. Grayling

"[The argument] that only God can underride objective values was refuted by Plato in 390 B.C. in an argument that he gave in the first and simplest of his dialogues, The Euthyphro. . . . What part of no [sic] Euthyphro don't you understand?"[114] – Alex Rosenberg

---

[112] Kitcher, *Life after Faith*, 27.
[113] Grayling, *The God Argument*, 157-8.
[114] Rosenberg, "Is Faith in God Reasonable?"

Response:
> "So long as a transcendent God is enforcing morality, the source of morality is irrelevant to our situation."

Alternative response:
> "Why does it logically follow that, if God determines what is good, that He is not all-good or all-powerful?"

Why it works:

In proper usage, the Euthyphro Dilemma, at best, establishes that God is either not all-good or He is not all-powerful. If God identifies what is good, and then enforces it, then goodness is separate from God. He is not all-powerful, because He did not create the good. If God creates the good by fiat, then He is capricious, because He can change His mind, and we're at the mercy of His arbitrary definitions of good. The argument is often (and errantly) extended to a second conclusion. If God is not all-powerful, He's not really God. The first option (which we haven't offered here as a response), is to reject the given dichotomy, in favor of a third option: God literally *is* good. He doesn't create or decide what good is, but His nature literally *is* good. Thus, the idea of good is not separable from God Himself.

However, we opt for simply accepting the false premise that's been offered, because nothing useful follows. Suppose God identifies a good external to Himself. So what? He's a supreme power who enforces it, and so it does nothing to change our situation. And, God is still necessary for us to recognize what good is. Alternatively, if God remains omnipotent, and declares what is good based on His own subjective preference, this doesn't limit His power, nor the fact that He has enforced what is good. Nothing that can harm the God of the Bible logically follows. To use a humorous quote from a coffee mug, "Me boss. You not." In this case, "He boss. We not." Perhaps the most fundamental—yet avoided—aspect of the discussion of God is that, if God exists, He gets to make the rules and we don't.

Argument:
"What does a Christian say at the bed of a dying child?"

Response:
"What does an atheist say?"

Why it works:
The most common usage of this argument stems from an argument made by the late Bertrand Russell (arguably the most influential atheistic philosopher of the last century). We don't offer examples here, because the argument usually takes the form of the classic "theodicy" (i.e., the problem of reconciling a good God with the existence of evil and suffering), and there are myriad ways it comes up in conversation. The first important observation is that, while atheists can attempt to make claims about morality, they are in no position to do so. For the atheist, there is no transcendent good or evil. These are ideas formed in isolation or in group-think by humans. Nothing more. Evil is not an objective reality for the atheist. Instead, there are only socially contrived or evolved preferences that will forever remain subjective. So, there can be no real complaint about the unfairness of the premature death of a child for the atheist. It is not really good or bad, right or wrong, but simply is. As Richard Dawkins once put it, "Nature is not cruel, only pitilessly indifferent. This is one of the hardest lessons for humans to learn. We cannot admit that things might be neither good nor evil, neither cruel nor kind, but simply callous—indifferent to all suffering, lacking all purpose."[115] For a cat to play with a mouse before devouring it is not *really* morally wrong. It just is. For male baboons of one tribe to murder a male from another is not wrong. It just is.

Further, to be an atheist is to have no belief in a deity, or the possible metaphysical implications that arise from such a being. Specifically, there cannot possibly be any claim to something like an afterlife, and many atheists are rather proud to assert that there isn't one. The point here is simple: the atheist has no conceivable means of giving hope to a dying child (or anyone else). The late

---

[115] Dawkins, *River out of Eden*, 112.

science historian William Provine made clear the grim reality of atheism on Ben Stein's 2004 documentary, *Expelled*:

> "It starts by giving up an active Deity, then it gives up the hope that there's any life after death. When you give those two up, the rest of it follows fairly easily. You give up the hope that there's an immanent morality. And, finally, there's no human free will. If you believe in evolution, you can't hope for there being any free will. There's no hope whatsoever of there being any deep meaning in life: We live, we die. We're absolutely gone when we die."[116]

If one's life consists only of pain and torment, then the atheist can offer them no other words of consolation than these: "everyone dies", "sorry about your luck".

In contrast, the Christian faith offers rectification in a very significant way. Believers are promised resurrection from the dead, and everlasting life (1 Thes. 4:16-17, Jn. 3:16), the reception of perfected bodies (1 Cor. 15:42-49), and that genuine justice will prevail when God judges every individual on a personal basis (2 Cor. 5:10). Additionally, the concepts of Satan's fall from heaven (Lk. 10:18), the Fall of man (Gen. 3), and inherited sin (Rom. 5:12) provide logical explanations for the suffering and evil that occurs within our world. Whether or not someone believes these teachings is not the point in the present discussion. The point is that the Christian worldview (and a number of other religions) actually offers an authentic explanation for human suffering, as well as a vision of hope for those enduring difficult times. The atheist's perspective cannot conceivably do either of those things; we live, we die, and we're absolutely gone.

---

[116] Fankowski, *Expelled*.

Argument:
"Can God make a rock so heavy He can't lift it?" (and similar so-called "problems of omnipotence")

Examples:
"[I]s God capable of acting in a way that would limit himself, such as by making himself not God, or making someone else God, or creating a challenge that he can't meet (like creating a stone that he cannot lift)? If he is, then there will be something he cannot do as a result of his action. If he is not capable of performing these kinds of actions, then, again, there is something he cannot do. So either way, God's power is limited and he is not omnipotent."[117] – Matthew S. McCormick

Response:
"That would depend on which 'God' you are referring to."

Alternative response:
"Can my answer to your question be both right and wrong at the same time?"

Why it works:

The first response demonstrates that the objection being raised fails because it assumes a form of "god" that deism cannot speak to, and to which traditional theism doesn't hold. They have effectively created a straw man of God, and attempted to knock it down. As a side note, this is perhaps one of the most important debate tactics one can learn. All too often, the theist is overly eager to defend their understanding of God. In some ways, this is like beginning to play defense on your own goal line. Do not abandon the rest of the field! Make them earn every inch of the god (little "g") argument. As a clear example, we encourage believers to revisit Ken Ham's debate with Bill Nye (even if you disagree with both of their views). Ken Ham allows Bill Nye to at-

---

[117] McCormick, "There is no God."

tack the existence of any god by simply attacking Ken Ham's conception of the Judeo-Christian God. But, does it logically follow that, if the God of Christian literalism is false, god does not exist? Of course not. Before you go letting them attack the attributes or descriptions of a specific view of God, make them deal with the possibility that any god might exist. Lesson over. Continuing . . .

The notion that God must be capable of logical contradictions is also a straw man. Just as we can confidently say that an argument cannot be true and false at the same time, we are also well within our rights to say that God can neither contradict His character nor accomplish acts that are outside the realm of logically coherent possibilities. For example, if God could create a square circle, then it wouldn't be a square, it would be a circle. We need not defend logical impossibilities. Put another way, theists do not (or should not) claim that God is able to do literally *everything* imaginable. To use the claims of Christianity for a moment, God cannot become incarnate in the man Jesus and not become incarnate in the man Jesus, or be Himself and be Satan, or love sin and hate sin, and so forth. As these examples demonstrate, God cannot violate His own identity in principle or in any action He may take. Thus, answering "no" to such omnipotence arguments is not a defeater of God.

Argument:
"If God exists, He is either incompetent or malicious." (a.k.a. the argument from poor design.)

Examples:
"There are many other aspects of biological life that astonish with their clumsiness, half-assedness, inefficiency, pointless superfluities, glaring omissions, laughable failures, 'fixed that for you' kluges and jury-rigs, and appalling, mind-numbing brutality."[118] - Greta Christina

---

[118] Christina, "Truth about Science vs. Religion."

"It isn't simply that these designers are imperfect, but that a truly intelligent designer, liberated from any constraint to produce descendants from previous organisms, would be expected to do much better."[119] – Philip Kitcher

"I think Intelligent Design implications are blasphemous, because they imply that God is inept. Like in the design of the jaw. And everything else... every animal or plant is incompetently designed, and is a cruelty.... I prefer to see this as natural selection, rather than [as] a consequence of design by an intelligent designer, the Creator.... I don't want the God of benevolence and the omnipotent God to be given the credit for having made that creation."[120] – Francisco Ayala

Response:

"When a major automaker issues a recall on faulty parts, or gets negative reviews on the layout of the interior, do you believe their car was not designed by intelligent agents?"

Why it works:

There is a reason that we have placed this argument against ID (and Creation Science as well) here at the transition from philosophy to theology, rather than in the science section. If you get to this point in a debate, you've already won for all intents and purposes. The argument from poor design is the last ditch effort to destroy the claim of intelligent causation in biological systems. But, it does not try to do so by any scientific argument. Rather, it appeals to a particular theological conception of God. Thus, it accepts the possibility of an intelligent designer that is less-than God, and demands a particular expectation of what God ought to do. So, before entertaining the claim, be sure to mention that you are now leaving the realm of science, and moving into theology. To quote William Lane Craig, responding to the idea that the de-

---

[119] Kitcher, *Life After Faith*, 48-49.
[120] Ayala, "Is Intelligent Design Viable?"

signer of life was more like a vengeful Greek God than the biblical Yahweh, "Zeus will do." That is, a bad designer is still profoundly different than no designer at all.

The argument from poor design commits several other false steps. First, it presumes that we have the knowledge sufficient to know what a better or worse design is. This would itself require god-like understanding of all functionally viable alternatives. Second, it assumes that a perfect design, as defined by us, is logically possible. Yet, von Neumann and Morgenstern (1947) showed long ago that it is mathematically impossible to maximize more than one variable in a system simultaneously.[121] People in the field of engineering refer to this as "constrained optimization". For example, one might point to the circuitous route of an artery in a giraffe, or charge that the cheetah should be still faster than it already is, and that it shouldn't get exhausted as a sprinter. Yet, there is no reason to think these variables could (in principle) be maximized simultaneously. Third, it denies any theological consideration of either flexibility or evolvability in created things. It is completely consistent (and posited by every view of theism) that life forms can change, adapt, evolve, or even devolve within some constrained range. This is particularly true for those who see a full fall of creation with the Fall of man. Genesis says as much for many believers. Fourth, if satisfied, the argument leads to the expectation of a static world, in which no organisms can evolve to be any better than they already are, because each organism is perfect from the beginning (regardless of later changes to the environment). None of these assumptions are consistent with the Christian understanding of God. Thus, we have yet another straw man.

---

[121] Neumann and Morgenstern, *Theory of Games*.

Chapter 4: Arguments from within Theism

Argument:
"You just feel there must be a conflict between science and faith. Theistic evolution sees no conflict between them."

Examples:
"Splitting faith from reason and pitting science against religion. . . . God is the author of all truth, both religious and scientific, so there can be no disjunction between them unless truth is pitted against itself."[122] – Cynthia Crysdale and Neil Ormeror

"[W]e have no discomfort with mainstream science. Natural selection as described by Charles Darwin is not contrary to theism. Similarly, we are content to let modern evolutionary biology inform us about the mechanisms of creation with the full realization that all that has happened occurs through God's activity."[123] – BioLogos website

"If two ideas are not in conflict, they have no need of reconciliation."[124] – Kenneth Miller

Response:
"Did Jesus rise from the dead?"

Why it works:

While the view is growing in popularity, the intellectual case for theistic evolution has never worked. Frankly, it is doomed to be logically incoherent, and the defeater for the view is rather easy. Here, we opt to use Jesus Christ as the stumbling block for the Christian theistic evolutionist. The claim of theistic evolution is that purely scientific (which is to say, naturalistic) explanations for the unfolding of the universe's history are somehow compatible with the God of Christianity. That is, the theistic evolutionist

---

[122] Crysdale and Ormeror, *Creator God, Evolving World*, 17.
[123] BioLogos, *BioLogos Different*, Lines 45–7.
[124] Miller, *Ken Miller Weighs in*, Lines 68–9.

embraces the proposed full efficacy of purely naturalistic explanations for phenomena.[125] They don't feel a need to invoke the direct action of God anywhere in the history of the universe, nor for life on Earth.[126] So, rather than debate cosmology or evolution, it's easier to just ask if they attest to the birth, death and resurrection of Jesus Christ. This is a lose-lose dilemma for the theistic evolutionist. If they deny the resurrection, then their Christianity is false. If they affirm it, then naturalism fails to be fully sufficient to explain history, or the nature of our species for that matter. It would follow that absolute loyalty to methodological naturalism is unwarranted, and things outside the methodology are perfectly rational. (Or, they admit that their belief in Jesus is irrational). Before further unpacking this move, consider a tacit admission of these facts by theistic evolutionist Darrel Falk:

> "Faith in the resurrection is thus the single most important belief that Christians hold. Is it scientifically credible? . . . It is not. Yet this is the position we hold. Furthermore, if we ever cease to hold it, then we have ceased to be Christians . . . I believe that there is good reason to accept that the Creator of all life visited this earth in the form of a human being. Given that belief, anything is possible, including the suspension of laws that he put in place to begin with. . . . Testing this claim is beyond the reach of the tools of science . . . to a scientist . . . the belief in a risen body is irrational."[127]

---

[125] I (WR) have written at length about this elsewhere, and will simply say that the generic theistic evolutionist is unwilling to invoke the direct action of God in manipulating the natural world from the birth of the universe to (at least) the rise of human civilization (thus also affirming the idea that we are evolved primates that did not have to emerge from the evolutionary process). Most question or outright reject miracles as well as God's willingness to engage petitionary prayer.

[126] For a sterling example, see page 200 of Francis Collins's book, *The Language of God*, in which he offers a full history of the universe from the Big Bang to the evolution of humanity. He fails to invoke God even as the cause for the universe, and doesn't mention God's direct activity until the supposed ensoulment of our species.

[127] Falk, *Coming to Peace with Science*, 210.

This is a self-impaling view. If "anything is possible, including the suspension of [natural] laws", then the theistic evolutionist has two profound problems: First, how could he or she ever believe that another view is actually wrong? Science is silent as to whether or not God created in six days, revealed Himself in a donkey and a burning bush, turned water into wine, or actively suspends His laws to produce miracles in present times. If science can't tell us about what really happens in the past, what good is it? And how would we even know if science was telling us what *really* happened, if our beliefs about history defy the very science we rely on? Second, what on earth are the "good reasons" for believing in such a suspension of natural law? After all, the justification of theistic evolution is that we must accept the primacy of natural law based on scientific evidence. So what other lines of evidence have rational primacy over science? They never talk about it. They choose, very literally, to stand side-by-side, shoulder-to-shoulder with atheists and attack all other forms of creationism, while at the same time picking and choosing their own places to make supernatural claims about reality. This is not missed by people like Jerry Coyne, who keeps records on the number of times that theistic evolutionists denounce and reject "god-of-the-gaps" arguments from other theists, and then immediately invoke their own god-of-the-gaps claims.[128] If the theistic evolutionist is going to open the door for the spiritual to act in the physical (often called "supervenience"), then they must justify where they think that door closes, and why. The theistic evolutionist would have to do so without reliance on science, which he or she would have already trumped in claiming Jesus's resurrection.

Assuming the theistic evolutionist affirms the resurrection of Jesus, it's fair to then press harder in two additional areas. First, if Jesus was a physical man, with a physical body, placed dead in a physical tomb, then physically absent from it, and physically appearing to many others after his death, then the methods used to detect physical events should be sufficient to detect miracles (since they take place in or on physical substrates). So the claim that science is "silent" with respect to miracles is false. If science

---

[128] Coyne, "God of the Gaps."

describes and explains physical reality, and miracles pervade into that reality, then the physical residue of their happenings are evidential (at least in principle). In fact, any "good reason" the theistic evolutionist has for belief in the resurrection of Jesus ultimately ends up being some type of physical evidence after all—an empty tomb, the records of Roman and Jewish historians, the accounts of the apostles (most notably, Paul's conversion and the other personal revelations of Jesus in Acts), or more recently, the mass sightings of Jesus in a predominately Muslim village days *prior* to the showing of the Jesus Film.[129] It would be foolish to argue that the personal experience of Jesus Christ moving internally in our hearts and minds would be sufficient evidence to deny the full continuity of natural law.[130] Most scientists feel these experiences also reduce to materialistic explanation. Further, if a personal spiritual (immaterial) revelation is sufficient, then why would we doubt the experiences of the various other religions, or of madmen who hear God's voice in their heads?

Argument:
    "God could've used evolution to create."

Examples:
    "We at BioLogos believe that God used the process of evolution to create all the life on earth today [and] accept the science of evolution."[131] – BioLogos website

    "[S]aying that mutation is random has nothing to do with whether evolutionary change is fulfilling God's purpose or unfolding according to God's design."[132] – Joan Roughgarden

---

[129] Thus, science would have a place in considering the veracity of metaphysical agency acting in the world.

[130] Let us be clear, we do not deny personal and private revelation. Only that it is insufficient to warrant rational belief, without being augmented by some form of objective evidence.

[131] BioLogos, *BioLogos Different*, Lines 1–4.

[132] Roughgarden, 47–8.

"Evolution, according to process theology, occurs in the first place only because God's power and action in relation to the world take the form of persuasive love rather than coercive force . . . to compel, after all, would be contrary to the very nature of love."[133] – John Haught

Response:

"Can you guarantee a particular winner using a chance-based lottery?" or "Can you be intended and unintended at the same time?"

Why it works:

Vague hand-waving at the hypothesis that God could use evolution to create is commonplace among theistic evolutionists. Their view relies on two premises, both of which must be demonstrated before the conclusion can obtain. First is the assumption that God is the Creator. Second is the belief that evolution describes creation. If true, it would follow necessarily that God is compatible with evolution. But, this greatly diminishes what the theist means by "create" and "intend". As John Polkinghorne describes it, "an evolutionary universe is theologically understood as creation allowed to make itself."[134] So, it's important to make the theistic evolutionist clearly articulate what they believe God was trying to accomplish by the process of cosmic and biological evolution. By their view, God is really not directing things toward any particular outcome or specified goal. He has set the universe into motion, and it is now creating itself.

A few other problems emerge. For example, if numerous chance-based or stochastic processes drive such a creation, then the products of those processes are not guaranteed and cannot be known ahead of time. God could not have intended for our species to evolve. Most theistic evolutionists admit this. Perhaps most transparent about it, Kenneth Miller proudly announces that, "mankind's appearance on this planet was not pre-ordained", and that we are, "an afterthought, a minor detail, a happenstance in a

---

[133] Haught, *God after Darwin*, 42.
[134] Polkinghorne, *Faith, Science, and Understanding*, 23.

history that might just as well have left us out."[135] There was no guarantee that our planet would produce life, nor that life would persist or become intelligent. So this line of argumentation really destroys God's ability to produce specified outcomes, and renders our existence accidental and unintended.

Without going too far into the discussion, it's also important to mention that, if evolution was God's desired method for creation, then He becomes morally responsible for a system that destroys more than it creates, and produces horrific suffering and misery in the form of developmental defects, malformations, and diseases. One would have to point to a stillborn set of twins, conjoined at the neck and sharing one heart, as a product of God's desired mechanism for creating. If one thinks long enough about it, the "God used evolution" mantra also eliminates any justification for the Adam and Eve characters, the Fall of man, or the corruption of creation that would necessitate His return and restoration of the creation. For the theistic evolutionist, there was no Adam (only "gradual polygenism", in which, "the Image of God and human sinfulness were gradually and mysteriously manifested across many generations of evolving ancestors"[136]), and thus there was no Fall or corruption of the creation. In fact, theistic evolutionists typically hold that the creation begins chaotic and evolves towards perfection (rather than being designed as "very good" and then falling into decay). If God used evolution alone to do His creating, He in no way resembles the God of Judaism and Christianity.

Rejoinder:
> "Given that God is omniscient, and has perfect knowledge of the past, present and future, what looks like chance to us is absolute surety from His view."

---

[135] Miller, *Finding Darwin's God*, 272.
[136] Lamoureux, *Accept Evolution*, 138. It's worth noting that the term "polygenism" is precisely the same term used by the Nazi's (as well as biologists like Ernst Haeckel) to distinguish between human races.

Response:

"So God still used a process of death, suffering, rivalry, and extinction to achieve His intended creation? Has the process stopped with us?"

Why it works:

This is a valid rejoinder in that, so long as we have good reason to believe that God exists and that He does in fact have complete knowledge of the future, what looks random to us can be guaranteed to Him. Theologically, it does not free the theistic evolutionist from the issues that arise with respect to God being the creator of evil and suffering, nor of the issue of free will (if God already knows that we will do a particular thing, then it is impossible that we could do otherwise, so we don't have free will in this sense, and He is ultimately responsible). Admittedly, these are theological considerations that exist with and without the assumption of evolutionary creation. However, this should limit the intellectually honest evolutionist. It would no longer be fair to suggest that evolution could've resulted in different intelligent creatures by different paths. Evolution becomes as fixed as every other feature of reality. God is now ordaining who wins the lottery and who is struck by lightning, or who gets cancer and who spontaneously heals from it.

Argument:

"Darwin is God's gift to religion."

Examples:

"The theory of evolution by natural selection is Darwin's gift to religion."[137] – Francisco Ayala

"Thank God for Evolution"[138] – Michael Dowd

"Can theology after Darwin, instead of retreating from or only reluctantly accommodating Darwinian

---

[137] Ayala, "Darwin's Gift to Science and Religion."
[138] This is the title of Dowd's 2009 book.

ideas, actually embrace them with enthusiasm?"[139] – John Haught

Response:
"So, a process driven by death is a gift from God?" or "Why then did Darwin's own theory lead him to abandon his faith in God?"

Why it works:

As Michael Antony wrote, "The problem with Darwin's mechanism of natural selection is its origins in a theory of Malthusian catastrophe. He saw evolution as driven by huge death rates . . ."[140] Thus, these almost-meaningless genuflections to the divine nature of biological evolution render down to serious theological problems with evil and suffering. The previously mentioned argument that a God who uses evolution is a God who is the wellspring of evil and suffering applies here as well. What "gift" offers an existence "red in tooth and claw", is characterized by the struggle to survive, produces more pain and suffering than joy and advance, and utilizes a mechanism that literally causes us to be sinful?[141]

A Partial Synthesis

It would be an act of pure hubris to suggest that the two of us, just neophytes in investigating the God Hypothesis, could offer anything like a synthesis on this matter. Surely, we would have no insights that have not been realized previously by someone else. Still, there does seem to be a general lack of holistic discussion in present-day public discourse. During debates and in most books, a series of arguments are trotted out, and we are left to evaluate each

---

[139] Haught, *God after Darwin*, 47. Incidentally, this quote appears in his chapter "Darwin's Gift to Theology."
[140] Antony, *The Masculine Century*, 161.
[141] As the theistic evolutionist Karl Giberson put it, "*sinfulness* is mainly *selfishness*. . . Selfishness, in fact, drives the evolutionary process . . . After many generations selfishness was so fully programmed in our genomes that it was a significant part of what we now call human nature." (*Saving Darwin*, p. 12).

one, and give it a "yes", "no" or "maybe" vote. At some level we are asked to tally up points on each side of the ledger, and declare a winner. But, arguments for or against the existence of God are not just overlapping; they augment and reinforce one another. Some points carry more weight than others. Additionally, the force of compounded arguments might be greater than the sum (or even the multiplicative product) of the individual parts. Periodically, it's important to stand back, take the wide-angle panorama of the totality of the discussion, and ask which view(s) pass(es) the epistemological sniff test. In this book, we have briefly touched on the arguments for and against God (or at least the necessity of a metaphysical being) across the broad areas of science, philosophy and even within theology itself. In closing, we would like to produce a formal step-wise argument for minimal ascription to metaphysical agency: deism.

As a debate tactic, we prescribe not taking the traditional route in defending the God Hypothesis. Most debates or discussions of God start at polar ends of the spectrum and attempt to win by hail-Mary. For example, why must we begin a defense of belief in God with some cosmological argument for God's existence? This is the grandest—and incidentally most scientifically moribund—place to start. Conversely, why begin a debate for the existence of God with a defense of the resurrection of Jesus Christ? Why open with a particular view of Christianity at all? It's not that good defenses can't be made for these things, but that they should be steps that proceed from more modest claims. We propose starting with things that are so clearly commonalities that they don't even qualify as arguments (even though they do work in favor of our case). There are certainly alternative strategies to making the case for God, but here is one possible path:

Step 1:
> "Do we agree that science, as an epistemology, accepts the existence of metaphysical realities?"

Why it works:
This argument can also be made as a declarative statement. Science's approach to "knowing" things is always described as an

evidence-based evaluation of competing hypotheses, all of which are restricted to naturalistic processes. Science is about facts, experiments, reproducibility, etcetera. Rarely mentioned are the assumptions made by science before it begins exploration of knowledge or truth. It turns out that there are many metaphysical realities that must be accepted before science can work. Numbers, like the integers or *pi*, are "abstract objects", which is to say that they have no physical properties like mass, volume, charge or spatio-temporal location. They are not part of the material universe (though they do relate to it), and thus are metaphysical. Yet, they are real in some sense. For example, if we take two apples and add them to a group that already contains two apples, we have four apples. This would be true whether or not any human could make such an observation. However, what if we attempt to subtract six apples from the group? There is no such thing as negative two apples. Negative two doesn't exist as a feature of material reality. Thus, we see that these mathematical objects are true in the abstract, but are not always applicable to concrete reality. Several other metaphysical assumptions, including mathematical truths and relationships (like $\sqrt{2} = 1.4142$, or $sin^2(t) + cos^2(t) = 1$) must also be assumed.

Further, science must assume that these abstract objects and truths exist *before* it can examine or explain any physical phenomenon. The entire enterprise of science depends on our ability to convert observations into "data" (quantified, itemized and categorized information), and science also relies on such mathematical truths in order to ascertain the "significance" of a pattern or finding. Frankly, you shouldn't get much push back from any atheist on this point. To quote the atheist philosopher Alex Rosenberg, "[We] adopt abstract objects—the objects of mathematics—as existing even though they are abstract, even though they are not concrete, even though they are not physical items in the world. Why? Because they [are] indispensable to the predictive power of science."[142]

Finally, logical truths must also be assumed. For example, if A = B, then B = A (called the "symmetry of identity"), or, things

---

[142] Rosenberg, "Is Faith in God Reasonable?"

like the law of non-contraction (A cannot = B and also not = B at the same time). Numerous logical laws must be true in order for science to work. Most notably, the "if x, then y" logical structure of scientific predictions must be true. Other more contentious examples of abstract objects are things like ideas or propositions (which have no physical properties, yet are assumed even as you read this sentence), or the capacity for a sentient being to have objective observation of a phenomenon (rather than itself being a caused result of prior phenomena). At any rate, this first step is both unavoidable for the atheist (even the nominalist must assume abstract objects), and it is incredibly damaging. It not only admits that non-physical realities exist, but binds such things to the applicability of science itself.

Step 2:
"Do we agree that there are things that science (in principle) cannot explain?"

Why it works:
Again, though you might hear claims like "science can (or will) explain everything", the one making such a statement doesn't actually believe it to be true. This is because there are many things that science is not capable of explaining, in principle. For example, anything that science pre-supposes in order to operate cannot be proven by science (such a claim would commit the fallacy of circular reasoning). It follows therefore that anything that is non-physical, metaphysical, or outside of the detectable universe is also unexplainable by science.[143] This is not a knock on science. The classic mathematical claim that $2 + 2 = 4$ is provable, but it is impossible to explain why such a truth exists. There are many more examples of things science cannot explain or prove, but these will

---

[143] For this reason, we have traditionally given way too much leniency to claims from the field of theoretical physics regarding realities outside of the known universe (for example, what causes the universe to come into existence, or what a multi-verse would be like). Abstractly, we might be able to use mathematics to postulate possible realities, but they can in no way be verified or proven by scientific methodologies. In light of this, we point to the growing list of physicists who are doubtful that things like multi-verse theories are even valid as "science."

suffice. What this step does is place hard constraints on the applicability of science. If someone argues for "omnipotent science" (as Peter Atkins once did in a debate), then we have castrated such a fantasy. There are truths and realities beyond the scope of science, and once again, science itself often relies upon such truths.

Step 3 (optional):
"For something to be 'science', must it conform to physical laws and processes, and be logically or rationally coherent? Further, must a scientific theory or concept explaining a phenomenon be demonstrable, testable, or experimentally repeatable?"

Why it works:
Going this route is likely to make for a more resistant (rather than agreeable) discussion. For this reason, one may move straight to step four. But, if you want to further delimit (or even expose) science, you can push a bit further on what qualifies as "science". Most people who argue against the existence of God love the simplification of science to facts, observable evidence, testability, reproducibility, and most importantly, rationality or reasonableness. However, science is not strictly bound to any of these (save for perhaps "facts"). Just to show a few places where these classical uses of the term "science" fail:

The most glaring example can be found in certain aspects of theoretical cosmology. Prior to the discovery of multiple lines of evidence suggesting that the universe began to exist (i.e., something like the Big Bang occurred), physicists had assumed that the universe was eternal, which is to say infinitely existing in the past. Shortly after accepting that the universe had a beginning, cosmologists began trying to explain how it came into being, and all of these explanations invoked either eternal physical laws or structures, or the equally logically impossible idea of self-causation (which again, suggests the universe is eternal). Because something must first exist before it can cause something else, self-causation is logically impossible. Also, the concept of infinity is, at best, only abstractly real, but not concretely existing. There is nothing in physical space-time that is infinite. Even in mathematics, it causes self-contradictory or nonsensical results.[144] Consider an analogy: can someone jump out of a bottomless pit? The answer is worse than "no". Even in an extremely deep pit, the individual could make some progress towards getting out, by simply jumping. But, if the pit has no bottom, the individual cannot get started, because there is no bottom to leap from. So it is with all claims of a past-eternal universe. Admitting that the universe did have a beginning, and then explaining that beginning by invoking a logically prior (not just temporally prior) physical cause only kicks the explanatory problem back one step. We then need to know how that physical system emerged. Worse, because it would lie outside of our physical universe, it would necessarily be impossible to directly detect, measure or demonstrate. Thus, any such explanation is not "science" in the traditional sense of the term, but would be some form of metaphysics. Finally, those metaphysical claims would not be applicable (or would be in opposition) to classical logic. The very same logic used to found the claim would be contradicted by it.

A second (and often related) example is that of the phenomena associated with quantum physics. A detailed understanding of quantum mechanics is not required to demonstrate that many interpretations of such phenomena are inconsistent with assumptions of logic. Take the old "Schrödinger's cat" example: We place a cat in a chamber containing a vial of hydrocyanic acid (which is both deadly and radioactive). A release switch on the vial will be tripped if a single atom of hydrocyanic acid decays (a quantum phenomenon). Because the quantum event of radioactive decay is literally actualized (i.e., is made to concretely happen or not happen) by the observer, the cat is both dead and alive until we actually observe it as being either dead or alive. Thus, we have a genuine violation of the law of non-contradiction, because A (the cat) is both B (dead) and not B (alive) at the same time. This can get extreme. For example, the late John Wheeler (one of the greatest physicists of the past century) was of the opinion that human ob-

---

[144] For example, summing all integers to infinity (i.e., $1 + 2 + 3 + 4 \ldots + \infty$) equals the number -0.08333. Does anyone actually believe that there is a point where an ever-growing value attained by summing positive integers magically becomes negative?

servations might well actualize the entire universe, both in the future and in the past![145] Because the location of a photon is actualized by an observer, the observation of a photon of light that left a distant star millions of years ago might allow an observer in the present to actualize millions of years of history instantaneously (i.e., that history is not actualized until the observer sees the photon). But of course, how do we arrive at the present situation in which an observer can actualize the photon's history, if there was not a concrete history producing it? At any rate, we must abandon fundamental logical assumptions in order to accept such a situation. Here, science again falls into what would classically be considered the irrational or illogical, in that the very logic used by science would be contradicted by it.

As we show below, another crucial area where strict definitions of science are abandoned is within the so-called "historical sciences". When you employ this critique of science, you are stating an unavoidable truth, but it is likely to meet objection in the form of incredulity. You're asking the individual to either give up many branches of science (because they no longer conform to the strict definition of science), or to allow that the definition of science is a moving target that cannot be used to exclude theories that lack direct experimentation or testability, or even those that run counter to our fundamental logical assumptions.

Step 4:

"Can science accommodate or detect intelligent agency?"

Why it works:

This too is a rhetorical question. Science can in fact detect the intentional acts of intelligent agents, and distinguish them from undirected or chance processes. Further, science is capable of identifying teleological intent in the material world. When an archaeologist is evaluating a sharp triangular stone, there are methods for concluding whether or not the stone formed naturally (by a

---

[145]Folger, "Does the Universe Exist if We're Not Looking?"

blind process like erosion or fragmentation) or was designed for a purpose by an intelligent agent (say, the forming of an arrowhead). As has been detailed elsewhere (most notably by William Dembski), numerous fields of science detect intelligent agency on a daily basis, including archaeology, forensic science, sociology, animal behavior, psychology, and many others. Most noteworthy among the list is the field of astrobiology, particularly as it focuses on the detection of extra-terrestrial signaling. The SETI institute believes that it can distinguish between background "noise" and intelligent patterning in the cosmos. So science is perfectly happy dealing with intelligent agency, meaning that the objection that science rejects agent-causation (in counter distinction to event-causation) is simply not valid.

So far we've offered no arguments that are in fundamental opposition to science. Yet, we've accomplished much. We've established the existence of metaphysical realities, limited science's explanatory power, and established a place for intelligent agency as a causal force that is detectable by science. This has hopefully created a space to "allow a divine foot in the door".[146]

Step 5:

"Given that abstract realities are not only permitted by science, but are actually assumed by it, *and* that science can accommodate the acts of intelligent agency, what would logically preclude science from at least considering non-physical agency as an explanatory cause?"

Why it works:

The question here is not whether there is evidence for transcendent intelligence, but rather whether or not we will consider the possibility of such a causal force. Even this step is not so far a leap. Abstract or metaphysical things (like numbers) can escape the causal closure of our universe. Yet, there are things (including the universe itself) that seem to require causes prior or external to the universe. However, things like abstract objects have no causal

---

[146] Lewontin, review of *The Demon-Haunted World*.

power. The existence of an abstract number or logical truth does not *cause* something to exist.[147] However, agents or minds are capable of acting as first causes, beginning chains of events without the need for prior causes.

As another baby-step towards belief in a transcendent being like God, consider one of the most enigmatic aspects of reality: it is comprehensible. Moreover, physical reality offers both transcendent and pervasive coherence. Why is it that, as we bore up or down in scale, from atoms to antelopes to astronomy, there are constant and intelligible laws that make sense both at those scales and the others above and below? Not only does coherence traverse scales of material organization (from the atomic to the cosmological), but the logical coherence of the universe also exists in a dramatic and lucid hierarchy of thought. For example, it is widely true that biology can be described in terms of chemistry, chemistry by physics, and physics by mathematics—which exists as abstract, logical sets of rules and relationships! That is, for some reason, abstract concepts and ideas (which are the properties of minds) can lay hold of, and explain, material reality via an arrangement that resembles a vast and immense set of Russian nesting dolls. We will again cite Paul Davies, who expressed this unexpected feature of reality, saying,

> "All scientists agree that doing science means figuring out what is going on in the world, what the universe is up to, what it's about. If it isn't about anything . . . you'd have no rational basis for believing that as you dug to deeper and deeper levels, what you'd uncover would be additional coherence and meaningful facts about the world."[148]

He went on to say, "Experience shows that as we go deeper and deeper into our enquiries into nature, we continue to find rational and meaningful order, rather than just a haphazard jumble of unrelated phenomena." So, we might ask ourselves a simple question: Given that there is no necessary expectation for such trans-

---

[147] Hat tip to William Lane Craig for championing this view for more than two decades.
[148] Davies, "Beyond Belief Conference: Session 5."

cendent coherence and comprehensibility, is this feature more or less likely given a transcendent creator? We believe that, on balance, the probability of such a state of affairs is far more likely given a creator than not.

While certainly not persuaded to believe so, many atheists have at least acknowledged that something like a deistic "prime mover" god could exist. As a surprising example, Richard Dawkins (perhaps the last person we would expect to acquiesce to deism) recently admitted, "You could possibly persuade me that there was some kind of creative force in the universe [and that] there was some kind of physical mathematical genius who created everything."[149] In fact, openness to deism is a concession many atheists and agnostics are quite willing to entertain. The physicist Paul Davies has been fairly transparent about his openness to deism, as has John David Barrow. And many scientists are in fact deists. In 1997, the journal *Nature* reported that 40 percent of scientists believed in a personal God. However, a large number additionally fall into the group "deist". While the number of theists has declined since that time (now roughly 33 percent), more than half of all scientists ascribe to some form of god (51 percent).[150] Einstein was a deist, as were many of the most radical philosophers and politicians of the Enlightenment.[151] So deism remains a serious view at the "grown up" table of discussion in science and philosophy.

Finally, know when to quit. If you can get to the point of an agreed upon agnosticism on this issue, we consider this a victory (it is a massive concession to abandon the belief that god—little "g"—does not exist, and open the door to god's existence as a logical possibility). Further, you have at least built a bridge from atheism and philosophical naturalism to deism. The bridge is self-evident and undeniable, even from within science, and it is not in danger of being overturned by future findings within science. There are good reasons to accept the claims we've made thus far, and we believe there *always* will be. The conversation that pro-

---

[149] Dawkins, "Has Science Buried God?"
[150] Masci, "Science and Belief."
[151] Voltaire, Thomas Jefferson, and arguably even Spinoza.

ceeds from here is just that; it's a conversation, not a debate. It will deal with more than facts and logic. It will deal with historical evidences, theological considerations, emotional attachments, personal experiences, and internal conditions of the heart. We actually believe that, while you "can't argue someone into heaven", you can convince people that belief in God is rational. The next step is to help them discover the personal need for a God that might really exist.

Appendix

Defining evolution

One of the more frustrating problems with the "science and faith" discussion is the way in which evolution has been presented to the public. The secular cognoscenti all treat it as a foregone conclusion that evolution is true, and skeptical U.S. citizenry are treated as stupid for not accepting it. Yet, very little time is spent actually discussing what evolution represents in modern science. First, it is important to point out that there is a reason that this has become the dividing line on atheism and theism. Evolution played an important role in Darwin's drift from Christianity to agnosticism. It allowed Richard Dawkins to be "an intellectually satisfied atheist." And, most atheists see Darwin's theory of evolution as the completion of the Copernican revolution. Naturalism had seemingly gained victories in every area *except* the biological world. Darwin is thought to have sealed the fate of this last bastion of theism as well. Plus, evolution hits home. It doesn't just render us small in a massive universe. It renders us primates, in a series of animal forms that no one intended.

Perhaps the largest problem with public (and even academic) discourse regarding the "fact of evolution" is the incessant conflating of the term. Even in seeing the way Christian academic-types discussed my (WR) recent book with students, the efforts reduced to nothing more than weak arguments from incredulity in feeling that the book could be excused out of hand, because it errantly rejects evolution. Of course, my book did not reject evolution. It clarified what we mean by the term and what we don't know about the process. When the secular biologist, freethinking skeptic or theistic evolutionist uses the term "evolution", it's often not clear what is meant by it. For example, *all* creation accounts adhere to some form of biological evolution (i.e., We're unaware of any theology contending that God specially created every organism, and that those organisms are damned to some kind of fixity). So, the fact of evolution, in this broadest sense, is in no way opposed to *any* form of theology. More frequently, what is being referenced by "evolution" is the pattern of evolution. That is to say (as we discussed under the "we have the fossils, we win" argument), the individual is attempting to beat the more conservative ("fundi") Christian over the head with the pattern of evolution. Essentially, because there are fossils, and they suggest both grand epochs of time and transitions in living forms over time, it follows that a purely naturalistic account trumps the creative power of God's direct hand. This of course is a *non sequitur*. The fossil record does not clearly demonstrate a particular mechanism for biological evolution, nor even a convincing case for universal common ancestry. Much is being assumed—but not demonstrated—by such usage of the term. Again, the pattern is not apparently in conflict with most forms of theology (removing the time component, in the case of Young Earth Creationism, makes it generalizable to any and all theology). So why then are we to be compelled by the evidence to abandon conceptions of an active deity? Because, while the opponent will often reference the pattern of evolution, what they are really pilfering in is adherence to the Darwinian account. That is, when they say evolution, they *could* mean lots of things, most of which are not adversarial to theism. What they *actually* mean is a very specific mechanism that completely eliminates the direct action of God from consideration. But that mechanism must be named and demonstrated.

Prior to Darwin, there were conceptions of evolution. But Darwin was the first to propose a viable mechanism for explaining how evolution took place. At the time, he knew almost nothing of the genetic component of inheritance (though he and Mendel were contemporaries, Mendel's work was not broadly known and accepted until the beginning of the 20th century). But, Darwin did believe that some kind of heritable unit was responsible for passing traits on to progeny. His mechanism (pangenesis) was flatly mistaken. His general model, however, has seemingly stood the test of time: heritable variation + natural selection = change over time. That is, he believed that all life was derived from a common ancestor, and that his theory could explain the origins of new species (as well as the extinction of others).

But there has been a problem. Darwin's original theory was wrong. The most glaring evidence against it has been a fossil record (and a genetics) that shows a discontinuous pattern, with jumps and spurts in the diversification of life, amid long periods of stasis (a pattern biologists generally refer to as "punctuated equilibrium"). Darwin's theory depended on small changes gradually accruing over long epochs of time. He even offered a possible refutation of his thesis, saying, "If it could be demonstrated that any complex organ existed, which could not possibly have been formed by numerous, successive, slight modifications, my theory would absolutely break down."[152] Thus, he predicted a gradual increase in the differences between related species. The evidence has clearly demonstrated a pattern much to the contrary of Darwin's expectation. Yet, the love affair with Darwin lives on.

Today, modern evolutionary theory has become what we (BR and WR) often refer to as a blob. It seeks to explain why some organisms change gradually, while others appear on the scene drastically different, fully developed, with no clear set of transitional species to link them to ancestral types. It seeks to explain why some species don't seem to change at all over eons of time, and why others *de*volve, destroying the complexity they once possessed. When the gradual accumulation of point mutations fails, a smorgasbord of new mechanisms, expanding in all directions, are offered in order to salvage a "theory". Some argue for "front-loading" of first life, such that it can navigate through the sea of mutational space, and find adaptive solutions more easily (the mechanism for which remains completely missing).[153] In place of impotent point mutations, we now have "evo-devo" (evolution and development), in which mutations to regulatory genes (or extragenic sequences) allow for re-tooling of entire developmental pathways and products without requiring novel genes. Other drastic infusions of raw materials for selection to work on now come in the form of gene, chromosome and even genome duplication events, after which duplicate genes might be co-opted for new functions, without the organism giving up previous ones. If these sources of variation are insufficient, there is lateral gene transfer (which is no longer constrained to prokaryotes), viral transposi-

---

[152] Darwin, *Origin of Species*, 189.
[153] See Gerhardt and Kirschner's "facilitated adaptation", or the views recently offered by James Shapiro, as a couple of quick examples.

tion, and many other mechanisms for cross-species exchanges of DNA. There is talk of "emergent" complexity and "self-organization" of cells and cellular systems.[154] Thought to have been replaced by Darwin's original theory, Lamarkism is now back, revived by epigenetic scenarios where non-DNA molecules from parent organisms can influence development and gene expression in progeny.[155] There's also somatic selection, multi-level selection, top-down organizational theory, genetic drift, nearly-neutral models, and symbiogenesis, just to name a few.

We don't need to somehow offer a refutation of the Darwinian theory of evolution; the scientific community itself has long criticized the theory, noting that it is insufficient as an explanation.[156] If Darwin's theory was sufficient, we would have no need for the conga line of supplemental models now in circulation, most of which are more complex and less demonstrable than the ones they seek to supplement. The arc of this movement towards ever more mechanisms and head-spinning complexity suggests that the problem is getting worse. Biologists now effectively seek to overfit the data with their models, trying to account for anything and everything that they encounter. They must do so, because "evolutionary theory" *must* be correct—even if it involves cockamamie theories of primordial cells front-loaded with a guiding algorithm for adaptive evolution, given to them by space aliens! The biologists perform such feats of pretzel-like mental contortion to avoid the concern that methodological naturalism might be failing to crack this particular nut.

The real reason that evolutionary theory is a poor scientific theory is that it now fails to make itself falsifiable. While criticiz-

---

[154] Such theories rely on putative (but unverified) "emergent phenomena". Stuart Kaufman, among others, is so confident in this form of evolution, that he has essentially abandoned any possibility of the Darwinian view being a valid explanation for life's complexity, writing, "A fundamental implication for biological evolution itself may be that selection is not powerful enough to avoid the generic self-organized properties of complex regulatory systems persistently 'scrambled' by mutations." ("Emergent Properties in Random Complex Automata," 1984).

[155] Hughes, "Epigenetics."

[156] I (WR) summarize many of the most relevant examples in chapter six of my book *Shadow of Oz*.

ing Marxism, Karl Popper was right when he observed that, "every 'good' scientific theory is a prohibition: it forbids certain things to happen. The more a theory forbids, the better it is. A theory which is not refutable by any conceivable event is non-scientific. Irrefutability is not a virtue of a theory (as people often think) but a vice."[157] If true, the neo-Darwinian theory fails as a scientific theory entirely. David Berlinski's appraisal of evolutionary theory is that, "no theory has been confirmed since every possibility has been justified,"[158] and he's correct. Consequently, Darwinian evolution lacks any clear or generalizable predictive power about the future or the past. When somebody argues that "evolution is a fact", it is no longer even clear what they're referring to.[159] What is evolution? Which version of the theory are they invoking?

The Kalam Cosmological Argument

Most of what we've done in this book is based on *defenses* of faith claims against various forms of argumentation. But, many apologists (across the religious spectrum) have attempted to go on the *offensive* in producing arguments in favor of the existence of god (or some higher power). One of the more popular arguments is the Kalam Cosmological Argument (KCA), which can be traced to ancient Greek thought, but was more formally applied during the Moslem high age of science (roughly the thirteenth century). The argument is typically offered as follows:

1. All things that begin to exist have a cause.

2. The universe began to exist.

3. Therefore, the universe has a cause.

---

[157] Popper, *Conjectures and Refutations*, 34–7.
[158] Berlinski, *Devil's Delusion*, 188.
[159] Even the concept of a universal common ancestor is eroding in light of modern research. Not that we will suddenly find a convincing argument for extensive special creation (at least not as the landscape looks today), but that it is less and less likely that the assumption of a complete coalescence of all life is actually true (rather than an assumption being forced on data).

Notice that the core premises and conclusion do not directly make a case for God. They can be (and largely are) agreed upon by secular cosmologists and physicists. It's fair to add an additional stipulation that the cause of the universe cannot be itself (self-causation represents a logical contradiction, because something must exist before it can cause anything). Still, some philosophers retain self-causation as a live possibility. When they do, they necessarily make an argument from absurdity, and logic would no longer apply to the discussion. Alternatively, some will object that the universe could be an eternal flux of alternating expansion and contraction, and thus would require no cause (things that do not begin to exist do not require causes). This view is largely rejected by modern science. A more nuanced alternative is the idea that time does not exist external to the universe, and, while the universe may have a causal beginning, it does not have a temporal one. We are not convinced that any cause-effect relationship (roughly meaning an event or state change) can occur atemporally (without time). For the sake of addressing the KCA—which assumes the universe began to exist—we will accept the assumption that the universe did begin to exist in the finite past.

Assuming the first part of the argument, we have heard many jump from "Therefore, the universe has a cause," to "And that cause must be God." That is, the argument is augmented with an additional claim that the cause of the universe must be transcendent and have the characteristics of something like a god. We think this is a much harder case to defend, and therefore would encourage the use of alternative arguments in favor of God (see our section on arguing for minimal deism from science, titled "A partial synthesis"). One of the KCA's strongest proponents, William Lane Craig, has recently fleshed it out more thoroughly, so as to provide some means of getting from the cause of the universe to that cause being God. He argues that, " . . . the universe has a cause. And since the universe can't cause itself, its cause must be beyond the space-time universe. It must be spaceless, timeless, immaterial, uncaused, and unimaginably powerful. Much like . . . God."[160] He has expanded upon this conclusion elsewhere,[161] in

---

[160] Craig, "The Kalam Cosmological Argument."

which he assumes that, "there is a cause of the beginning of the universe and that this cause is itself uncaused, beginningless, and enormously powerful." Here, he assumes that any event is a change, and events connote time. Therefore, any cause beyond the space-time of our universe must be changeless and timeless. Our concern here is that such aspects of a cause for the universe are unwarranted, and seem to be derived *from* Craig's conception of God, not arguments leading *to* his conception of God.

It is beyond the scope of this book, but it is possible to believe in God (or even just god) without requiring that being to be spaceless, timeless or immaterial. As far as we can tell, we don't *necessarily* need any such attributes in the cause of our universe.[162] That is, it doesn't clearly follow that the cause of our universe must be timeless, spaceless, immaterial, or uncaused.[163] Why assume that there are no events or substance-based realities prior to or outside of the universe? It seems, to some degree, that Lawrence Krauss was correct when he told Craig, "we could put anything back there." Many alternative mechanisms can be offered as causes for the universe, and they need not be absolutely timeless, dimensionless, unchanging, etc. (furthermore, neither does God). They simply need to be external to the universe and have sufficient causal power. Thus, the Kalam ceases to be a definitive knock-down if mechanisms other than God can satisfy the need for a cause to the universe. This would be even more the case if that mechanism is presumed to be timeless or eternal. So, in final analysis, we are not arguing that the KCA (nor even its add-ons) are false. We are arguing that it seems not to offer a clear path to demonstrating the

---

[161] Craig, "Is the Cause of the Universe an Uncaused, Personal Creator?"

[162] Additionally, quantum entanglement allows for cause-effect realities that have no time component in our universe. So Craig's assumption that events require time may be false. During a recent podcast, Craig stated as much, conceding that "philosophers of time are pretty much split 50:50 on whether time is tensed or tenseless." ("Recent Responses to the Kalam Argument").

[163] We must clarify that we are not saying that the universe could be caused by something that is made of matter in the way we understand matter (i.e., as understood from within the universe's physical realities), but that it could still be substantive in that it could possess dimensional attributes such that it is neither metaphysically represented as "void" nor is it necessarily omnipresent (which is really how Craig seems to be using the term).

necessity of God. The idea that any cause external to the universe must be God is central to a related argument—Leibniz's Principle of Sufficient Reason—which we will now address.

To paraphrase, Leibniz once argued in rhetorical fashion, why is there something rather than nothing? His *On the Ultimate Origination of Things* (1697) offers an argument much like the KCA in that it apprehends the problem of the infinite regress fallacy. But, it is often seen as stronger than the KCA in that it readily accommodates the positing of an eternal universe (or universe-causing mechanism). Leibniz's interest was not in pinning God as a cause for the universe, but rather, as the ultimate cause for all existence—even if those things exist eternally. As he put it, "Hence it is evident that even by supposing the world to be eternal, the recourse to an ultimate cause of the universe beyond this world, that is, to God, cannot be avoided." In this sense, he grants the possibility of an infinite series or succession of events, but holds that we must logically demand a sufficient and ultimate reason for such a series, and argues that such a reason "can only be a necessary being which has in itself the reason for its existence." We (BR & WR) have previously considered this to be a profoundly powerful argument in favor of God. However, we have more recently begun to question the dichotomy presented by Leibniz. The argument depends on the logical possibility of nothing existing. This side of the dichotomy is not self-evident (and may be logically impossible). Going a step further, Leibniz has made the concession that we could have a situation where nothing exists, and this would mean that God (obviously, being "something") would therefore not exist. But, could God exist alone in a void of empty nothingness? Many theologians seem to think so. That is, absent the creation of the universe (or anything else), there would simply be the One Ultimate Cause amid utter nothingness. In this way, the argument seems to reduce to "If God, then God", which is a tautological truism. Further, if true, it negates Leibniz's central thesis, because it is possible for nothing to exist—instead of something—even if God exists. Thus, both nothing and something exist given God. The only alternative seems to be claiming that both creation and the Creator are timeless (i.e. have always existed). It's not fashionable to do so from within the faith, but we must acknowledge the atheistic claim "why assume nothing as the default?"[164]

Finally, associated with both the KCA and the Leibnizian claim of sufficient reason, is the so-called Anthropic Principle argument (which is essentially identical to the Cosmological Fine-Tuning argument), which holds that the precision seen in the fundamental laws of the universe are so improbable that they demand an intelligent designer. The fine-tuning of the universe is indeed spectacular, and it troubles many atheists.[165] Many feel that this argument does not rely on what we don't—perhaps even can't—know, but rather on what we *do* know. For example, among the ever-growing list of finely-tuned values in the laws governing our universe is the cosmological constant, which is tuned to a precision of a decimal value that has one hundred and twenty zeros in front of it. Change the value in an infinitesimal way and the universe may enter collapse or hyper-expansion. Atheists and theists alike largely agree on this (and similar) facts. However, we again caution usage of the argument. Precision does not necessarily equate to probability. For example, if you discovered that the ambient temperature of your home was 72° F, this would not look very fine-tuned. But if we had sufficient means for measuring the temperature of your home with much greater precision, we might find that the actual temperature is 72.000000000001° F. The temperature exists at a value with a precision of one in a hundred billionth. But does that make it less probable? Not necessarily. It only makes the value more precise. The only way we could make an argument for the improbability of the value would be to know the relative probability of other temperatures. When applied to the universe, we do not know what the actual probability of any other setting for a constant might be. Because we cannot (as least so far as we're aware) manipulate the values of physical laws, we cannot establish the probability of any alternative tuning of these constants. Another liability of the argument is that is supposes that

---

[164] This argument is made by Peter Boghossian in his book *A Manual for Creating Atheists* (p. 149).
[165] Several have conceded that it is the best argument for theism, and fine-tuning was credited as playing a large part in the conversion of the late Antony Flew to belief in God.

finer-tuning of constants and values increases the probability that God exists. That is to say, if values for constants were less precise, this would apparently be evidence against the existence of God. Aside from offering us a double-edged sword, such a feature of fine-tuning arguments also contradicts the previously mentioned argument from Leibniz, in that Leibniz would look at the existence of any universe (no matter how simple) as evidence for God.

To be clear, we are not saying that any of our concerns render these arguments powerless or even counter to the existence of God. At minimum, we feel that users should be careful in applying them, and not over-reach with respect to the conclusions that can be drawn from them. We fully acknowledge that proponents of these arguments will likely offer rejoinders, and that we may be ignorant of important nuances in making them. If any of the rejoinders satisfy our concerns, then we would promote their usage.

Bibliography

Alexander, Denis. "Intelligent Design is Not Science". *The Guardian* (December 3, 2009)

Anderson, Eric. "Probability & Design". *ID the Future Podcast* (June 6, 2015).

Antony, Michael. The Masculine Century: A Heretical History of Our Time. iUniverse, 2008.

Atkins, Peter. "What is the Evidence for/against the Existence of God?" Carter Presidential Center, Atlanta, GA, (April 3, 1998)

Ayala, Francisco. "Is Intelligent Design Viable?". Indiana University, Bloomington, Indiana, (November 5, 2009).

———. 2009. Evolution by Natural Selection: Darwin's Gift to Science and Religion. *Theology and Science*, 7(4): 323-35.

———. *Darwin's Gift: to Science and Religion*. Washington DC, USA: Joseph Henry Press, 2007.

Ball, Philip. 2013. DNA: Celebrate the Unknowns. *Nature*, 496: 419-20.

Berlinski, David. *The Devil's Delusion: Atheism and its Scientific Pretensions*. New York: Basic, 2009.

Berry, Matthew. "100 fantasy football facts for 2015". *ESPN* (June 30, 2015) http://espn.go.com/fantasy/football/story/_/id/13101710/100-facts-2015-season-fantasy-football

BioLogos.org. "How is BioLogos Different from Evolutionism, Intelligent Design, and Creationism?" Online: http://biologos.org/questions/biologos-id-creationism.

Boghossian, Peter. *A Manual for Creating Atheists*. Charlottesville, VA: Pitchstone Publishing, 2013.

Branch, Glenn. "Intelligent Design is Not Science, and Should Not Join Evolution in the Classroom". *U.S. News* (February 2, 2009).

Carroll, Sean. "Does the Universe Need God?" http://preposterousuniverse.com/writings/dtung/

Carroll, Sean B. *Endless Forms Most Beautiful: The New Science of Evo Devo*. W.W. NY : Norton & Company, 2006.

Chamary, J.V. and Laurence D. Hurst. "How Trivial DNA Changes Can Hurt Health". *Scientific American*. (June, 2009)

Collins, Francis S. *The Language of God: A Scientist Presents Evidence for Belief*. New York: Simon & Schuster, 2006.

Cooper, Rob. "Forcing a religion on your children is as bad as child abuse, claims atheist professor Richard Dawkins". *UK Daily Mail* (April 22, 2013).

Cotton, John. "Intelligent Design Creationism Is Not Science". http://www.physics.smu.edu/pseudo/IntelligentDesign/

Coyne, Jerry. "God of the gaps—still with us." Posted March 16, 2013. https://whyevolutionistrue.wordpress.com/2013/03/16/god-of-the-gaps-still-with-us/

——. "How to Get Rid of Religion". https://whyevolutionistrue.wordpress.com/2012/11/05/how-to-get-rid-of-religion/

Craig, William L. "Recent Responses to the Kalam Argument". *Reasonable Faith Podcast*, November, 15, 2015.

―――."The Origins of Aggressive Atheism", Reasonable Faith Podcast (May 31, 2015).

―――. "Is the Cause of the Universe an Uncaused, Personal Creator of the Universe, who sans the Universe is Beginningless, Changeless, Immaterial, Timeless, Spaceless, and Enormously Powerful?" Posted October 10, 2010. http://www.reasonablefaith.org/is-the-cause-of-the-universe-an-uncaused-personal-creator-of-the-universe

―――. "Who Created God? A Response to Dawkins." Posted June 18, 2009. https://www.youtube.com/watch?v=C3rjboY2MVw

―――. "The Kalam Cosmological Argument". http://www.reasonablefaith.org/transcript-kalam-cosmological-argument

Crysdale, Cynthia and Ormerod, Neil. *Creator God, Evolving World*. Fortress Press, 2013.

Darwin, Charles R. *On the Origin of Species by Means of Natural Selection, or the Preservation of Favoured Races in the Struggle for Life*. 1859. Reprint. London: Arcturus, 2008.

Davies, Paul. The Cosmic Blueprint: New Discoveries in Nature's Creative Ability To Order the Universe. Simon & Schuster: New York, 1988.

Dawkins, Richard. *The God Delusion*. Mariner Books, 2008.

―――. "Has Science Buried God? A Debate with John Lennox". The Natural History Museum, Oxford University, (October 21, 2008).

―――. "Is Science a Religion?" *Skeptical Science* (re-post from *Humanist*) (January-February, 1997)

———. *River Out of Eden: A Darwinian View of Life.* Weidenfeld & Nicholson, 1995.

Dembski, William. *The Design Inference: Eliminating Chance through Small Probabilities.* Cambridge, UK: Cambridge University Press, 1998.

———, Behe, M., Miller, K. and Pennock, R. "Blind Evolution or Intelligent Design?" 2002.

D'Onofrio, David J.; Abel, David L. 2014. "Redundancy of the genetic code enables translational pausing". *Frontiers in Genetics,* 5:140.

Dowd, M. *Thank God for Evolution.* Plume Books, 2009.

Durston, Kirk. "An Essential Prediction of Darwinian Theory is Falsified by Information Degradation", July 9, 2015. http://www.evolutionnews.org/2015/07/an_essential_pr09 7521.html

Expelled: No Intelligence Allowed, directed by Nathan Fankowski. 2008. Salt lake City, UT: Premise, 2008. DVD.

Falk, Darrel R. *Coming to Peace with Science*: Bridging the Worlds Between Faith and Biology. InterVarsity Press, 2004.

Folger, Tim. "Does the Universe Exist if We're Not Looking?" *Discover Magazine,* June 1, 2002.

Frank, Adam and Marcelo Gleiser. "A Crisis at the Edge of Physics". *New York Times* (June 5, 2015)

Fu, Wenqing et al. "Analysis of 6,515 Exomes Reveals a Recent Origin of Most Human Protein-Coding Variants." *Nature* 10, no. 493 (January 2013): 216–220. doi:10.1038/nature11690.

Giberson, Karl W. *Saving Darwin: How to Be a Christian and Believe in Evolution*. New York: HarperOne, 2008.

Goldstein, Rebecca Newberger. "36 Arguments for the Existence of God". *Edge*, November 18, 2009. http://edge.org/conversation/36-arguments-for-the-existence-of-god

Grayling, A. C. *The God Argument: The Case against Religion and for Humanism*. Bloomsbury: New York, 2014.

Green, Emma. "The Origins of Aggressive Atheism". *The Atlantic* (November, 2013).

Greta, Christina. "The truth about science vs. religion: 4 reasons why intelligent design falls flat". Salon Magazine, August 4, 2014. http://www.salon.com/2014/08/04/the_truth_about_science_vs_religion_4_reasons_why_intelligent_design_falls_flat_partner/

Hall, David W., and Joseph, Sara B. "A High Frequency of Beneficial Mutations Across Multiple Fitness Components in *Saccharomyces cerevisiae*." *Genetics* 185, no. 4 (2010): 1397–1409.

Harman, Liz. "Scientist Puts Faith in Evolution". posted November 2, 2009. http://www.sandiego.edu/insideusd/?p=6476

Harris, Sam. *The Moral Landscape: How Science Can Determine Human Values*. Free Press, 2011.

Haught, John. *God after Darwin: A Theology of Evolution*. Boulder: Westview, 2000.

Hawking, Stephen W. *A Brief History of Time: From the Big Bang to Black Holes*. New York: Bantam, 1988.

Higgins, Penny. "Why 'Intelligent Design' (ID) is Not Science". *The Committee for Skeptical Inquiry* (March 1, 2006).

Hitchens, Christopher. *God Is Not Great: How Religion Poisons Everything.* Twelve Publishing, 2009.

Holt, Jim. "Why does the universe exist?" *Technology, Entertainment and Design (TED) 2014* (March, 2014).

Horvath, Anthony. *The Golden Rule of Epistemology and Other Essays.* Athanatos Christian Ministries Press: Greenwood, WI, 2015.

Hughes, Virginia. "Epigenetics: The Sins of the Father." *Nature* (March 5, 2014). Online: http://www.nature.com/news/epigenetics-the-sins-of-the-father-1.14816.

Hume, David. *An Enquiry Concerning Human Understanding.* Harvard Classics, Collier & Son, 1910.

Jeanson, Nathaniel T. 2013. Recent, Functionally Diverse Origin for Mitochondrial Genes from ~2700 Metazoan Species. *Answers Research Journal.* 6: 467-501.

Kauffman, S.A. 1984. Emergent Properties in Random Complex Automata. *Physica D: Nonlinear Phenomena*, 10: 145-56.

Kitcher, Philip. *Life After Faith: The Case for Secular Humanism.* Yale University Press, 2015.

———. *Living with Darwin: Evolution, Design, and the Future of Faith.* Oxford University Press, 2009.

Krauss, Lawrence. "Life, the Universe, and Nothing (III): Is It Reasonable to Believe There Is a God?". The City Bible Forum, Melbourne, Australia, (August 16, 2013).

———. "Life, the Universe, and Nothing (I): Has Science Buried God?". The City Bible Forum, Brisbane, Australia, (August 7, 2013).

Lamoureux, Denis O. *I Love Jesus & I Accept Evolution.* Eugene, OR: Wipf & Stock, 2009.

Lewontin, Richard C. Review of *The Demon-Haunted World: Science as a Candle in the Dark*, by Carl Sagan. *The New York Review of Books,* January 1997.

Li, G.W., Oh, E. and J.S. Weissman. 2012. "The anti-Shine-Dalgarno sequence drives translational pausing and codon choice in bacteria". Nature, 484(7395): 538-41.

Masci, David. "Scientists and Belief". *PewResearchCenter*, November, 5, 2009.

McCormick, M.S. "An Accumulation and Amplification of Doubts." Posted April 16, 2010. http://www.provingthenegative.com/2010/04/accumulation-and-amplification-of.html

Melott, Adrian L. "Opinion: Two Views of Intelligent Design". *Physics Today*, June 2002.

Meyer, Stephen C. *Signature in the Cell: DNA Evidence for Intelligent Design*. HarperCollins: NY, New York, 2009.

Miller, Kenneth, "How evolution is taught in the classroom", Host: Ira Flatow with Stephen Meyer, Kenneth Miller, Lawrence Krauss & Deborah Owens-Fink National Public Radio, November 8, 2002

———. "Biologist Ken Miller Weighs in on Scripture, Literalism and the Bible's 'Sexy Love Poem'". Online: http://www.godofevolution.com/biologist-ken-millerweighs-in-on-scripture-literalism-and-the-bibles-sexy-love-poem/.

———. *Finding Darwin's God: A Scientist's Search for Common Ground between God and Evolution*. New York: Harper-Collins, 1999.

Nagel, Thomas. *Mind and Cosmos: Why the Materialist Neo-Darwinian Conception of Nature is Almost Certainly False*. Oxford Press, 2012.

nssl.noaa.gov. "Frequently Asked Questions about Tornadoes" http://www.nssl.noaa.gov/education/svrwx101/tornadoes/faq/

Pennisi, Elizabeth. 2013. "The Man Who Bottled Evolution". *Science*, 342(6160): 790-793.

Polkinghorne, J. *Faith, Science, and Understanding*. New Haven: Yale University Press, 2000.

———. *Belief in God in an Age of Science*. New Haven: Yale University Press, 1998.

Popper, Karl. *Conjectures and Refutations*. New York: Basic, 1962.

Ramachandran, V.S. "Beyond Belief: Session Ten". The Salk Institute, La Jolla, CA, (November 7, 2006).

Roberts, Stephen F. "Quote History". accessed December 3, 2015. http://freelink.wildlink.com/quote_history.php

Roscoe, Lilly. "Stop Taking a Stand on Facebook – New Rules of Facebook #2". Posted on January 8, 2013. http://www.roscoelilly.org/stop-taking-a-stand-on-facebook-new-rules-of-facebook-2/

Rosenberg, Alex and Craig, William L. "Is Faith in God Reasonable?" February 1, 2013.

Rossiter, Wayne D. *Shadow of Oz: Theistic Evolution and the Absent God*. Wipf & Stock, 2015.

Roughgarden, Joan. *Evolution and Christian Faith: Reflections of an Evolution Biologist*. Washington, DC: Island, 2006.

Shermer, Michael. *Why Darwin Matters: The Case Against Intelligent Design*. Holt Paperbacks, 2007.

———. "ID Works in Mysterious Ways: Evolution Denial & Intelligent Design". *Huffington Post*, June 15, 2005. http://www.huffingtonpost.com/michael-shermer/id-works-in-mysterious-wa_b_2711.html

———. "Not Intelligent & Surely Not Science". (March, 2005) http://www.michaelshermer.com/2005/03/not-intelligent-not-science/

———, Michael, Prothero, Donald, Sternberg, Richard, and Stephen Meyer. "Has Evolutionary Theory Adequately Explained the Origins of Life?" Beverly Hills, November 30, 2009.

——— and William Dembski. Debate on Audio Martini podcast. December 7, 2005. http://www.michaelshermer.com/2005/12/william-dembski-debate/

Steinhardt, Paul. 2014. "Big Bang Blunder Bursts the Multiverse Bubble". *Nature*, 510(7503): 9.

Thoreau, Henry D. *Walden*. Empire Books, NY. 2013.

Tobin, Grant. "7.5 Million Americans Have 'Lost Their Religion' Since 2012". *Huffington Post* (March 13, 2015).

Truzzi, Marcello. 1978. "On the Extraordinary: An Attempt at Clarification". *Zetetic Scholar*, 1(1): 11.

von Neumann, J., and Morgenstern, O. *Theory of games and economic behavior*, 2nd ed. Princeton, New Jersey, Princeton University Press (1947).

Wen, Yan-Zi, et al. 2012. "Pseudogenes Are Not Pseudo Any More." *RNA Biology* 9, no. 1: 27–32.

Zaborske, J. M. et al.2004. "A Nutrient-Driven tRNA Modification Alters Translational Fidelity and Genome-wide Protein Coding across an Animal Genus". PLOS Biology, 12:12